SpringerBriefs in Education

For further volumes:
http://www.springer.com/series/8914

Mansoor Niaz · Marniev Luiggi

Facilitating Conceptual Change in Students' Understanding of the Periodic Table

 Springer

Mansoor Niaz
Marniev Luiggi
Epistemology of Science Group
Department of Chemistry
Universidad de Oriente
Cumaná
Sucre
Venezuela

ISSN 2211-1921   ISSN 2211-193X   (electronic)
ISBN 978-3-319-01085-4   ISBN 978-3-319-01086-1   (eBook)
DOI 10.1007/978-3-319-01086-1
Springer Cham Heidelberg New York Dordrecht London

Library of Congress Control Number: 2013942010

Printed on acid-free paper

Springer is part of Springer Science+Business Media (www.springer.com)

Acknowledgments

This work has been in preparation for many years and benefitted from the advice and support of our students, colleagues, and friends. Part of the work was supported by grants from Consejo de Investigación, Universidad de Oriente, Venezuela.

A major source of inspiration for this study was the seminal work of Stephen G. Brush (University of Maryland), based on his *The Reception of Mendeleev's Periodic Law in America and Britain*, published in 1996. Discussions with Michael Weisberg (University of Pennsylvania) were most helpful with respect to the role played by the atomic theory in Mendeleev's elaboration of the periodic table. Juan Quílez (IES Benicalap, Valencia, Spain) brought to our attention various historical sources with respect to the periodic table and also provided important feedback.

An anonymous reviewer provided constructive criticisms that helped to improve the monograph. We would like to express our sincere thanks to the following members of our research group, for providing criticisms and advice: Luis A. Montes, Ysmandi Páez, Arelys Maza, Cecilia Marcano, and Johhana Ospina.

A special word of thanks is due to Bernadette Ohmer, Publishing Editor at Springer (Dordrecht) and her staff (Marianna Pascale), for their support, coordination, and encouragement throughout the various stages of publication.

Contents

Abstract

Periodic table forms part of almost every high school and introductory university level chemistry textbook published on this planet. Despite its usefulness as a conceptual tool for organization of the chemical elements, understanding their properties, predicting new elements and a corrective device, most students consider it to be a difficult topic. The objectives of this study are: (1) Review of various aspects in the development of the periodic table based on a history and philosophy of science (HPS) perspective; and (2) Design of a teaching strategy based on the historical aspects in order to facilitate freshman students' conceptual understanding of the periodic table. Following historical aspects were considered to be important for the design of the teaching strategy: Role of the Karlsruhe Congress of 1860, accommodation of the chemical atoms in the periodic table, prediction of elements that were discovered later, corrections of the atomic weights, periodicity in the periodic table as a function of the atomic theory, and accommodation of argon in the periodic table. This study is based on two groups of freshman students enrolled in a Chemistry I course at a major university in Latin America. In order to avoid possible interactions, control group students received instruction in a semester prior to that of the experimental group. Both groups were asked to look for information on the periodic table from the Internet and traditional textbooks found in the university library. Following this, control group students discussed and solved problems found in the textbooks. Experimental group students were then exposed to the following phases of an experimental treatment: (a) Discussion of various aspects related to HPS; (b) Construction of concept maps; (c) Evaluation based on Posttest 1 (Items 1–4); (d) Classroom discussions based on students' responses on Posttest 1; (d) PowerPoint presentation by the instructor based on various HPS aspects; (e) Construction of new concept maps; (f) Discussion and comparison of the two sets of concept maps; (g) Evaluation based on Posttest 2 (Items 5–7); and (h) Five volunteer students participated in semi-structured interviews. Control group students were also evaluated on Posttests 1 and 2 and spent the same time in solving traditional problems, as the experimental group used to receive the treatment. Experimental group students provided conceptual responses on all items. Item 1 dealt with atomic theory as the criterion used by Mendeleev to order the elements, and 19 % of the students responded conceptually. Item 2 dealt with the relationship between the periodic

table and the early atomic theory, and 47 % responded conceptually. Item 3 dealt
with the question as to how Mendeleev could elaborate the periodic table before
the modern atomic theory, and 28 % responded conceptually. Item 4 asked if the
idea of ordering the elements originate with Mendeleev and 13 % responded
conceptually. Item 7 referred to periodicity as a function of the chemical atoms
(atomic theory) and 13 % responded conceptually. It was not expected that control
group students would respond conceptually. Nevertheless, one student on Item 1,
two students on Item 2, and one student on Item 3 responded conceptually. This is
an interesting finding and shows that given the opportunity to reflect even some
control group students can go beyond and improve understanding. Based on the
results obtained, a teaching strategy is designed for high school and freshman
students, which can facilitate students' conceptual understanding.

Keywords History and philosophy of science · Historical reconstruction ·
Periodic table · Dalton's atomic theory · Gay-Lussac · Avogadro · Karlsruhe
congress · Cannizaro · Newtonian method · Mendeleev · Meyer · Moseley ·
Periodicity · Accommodation · Prediction · Atomic mass (weight) · Atomic
number · Valence · Physicochemical properties · Correction of atomic weights ·
Noble gases · Modern atomic theory · General chemistry textbooks · Concept
maps · Conceptual understanding · Teaching strategy

Facilitating Conceptual Change in Students' Understanding of the Periodic Table

Introduction

The periodic table is considered to be an important topic of general chemistry courses in most parts of the world. Despite its usefulness as a conceptual tool for organization of the chemical elements, understanding their properties, predicting new elements and a corrective device, most students consider it to be a difficult topic. Furthermore, the long history of the development of the periodic table, starting from the early nineteenth century, makes any attempt at understanding the how and why of the changes, difficult to understand (Brito et al. 2005). Interestingly, Mendeleev's textbook (*Principles of Chemistry*, written between 1868 and 1870) was an endeavor to facilitate students' understanding of methods of observation, experimental facts, laws of chemistry, and perhaps most important of all, the "… unchangeable substratum underlying the various forms of matter" (Mendeleev 1897, Preface, p. vii). Indeed, this "unchangeable substratum" represents Mendeleev's fundamental presupposition with respect to the periodicity of properties in the periodic table as a function of the atomic theory (cf. Brush 1996; Niaz et al. 2004; van Spronsen 1969). In 1869, while presenting the first version of his periodic table, Mendeleev relied on the following important sources of information: Dalton's atomic theory, law of multiple proportions, Cannizaro's Karlsruhe Lecture, fairly reliable atomic weights (atomic mass according to modern terminology), atomicity (valence), and various physical and chemical properties of the elements (Weisberg 2007; Gordin 2004).

In writing his textbook, Mendeleev was particularly concerned to demonstrate to the students that chemistry is not simply a huge collection of facts that need to be memorized, but rather there is also "law and order" in this vast domain (cf. Bensaude-Vincent, 1986; Pattison Muir 1887). Furthermore, as early as 1874, a German language chemistry textbook had included Mendeleev's periodic table (Rammelsberg 1874; we are indebted to Stephen Brush for this reference).

Brito et al. (2005) analyzed 57 general chemistry textbooks (published in USA) and found that none described satisfactorily the role played by the atomic theory in understanding periodicity. To make matters worse, most textbooks emphasized

M. Niaz and M. Luiggi, *Facilitating Conceptual Change in Students' Understanding of the Periodic Table*, SpringerBriefs in Education, DOI: 10.1007/978-3-319-01086-1_1, © The Author(s) 2014

that the periodic table was primarily an inductive generalization and that Mendeleev had no theory or model to explain the periodicity of the elements. Furthermore, in general, textbooks give the impression that for almost 100 years (1820–1920), until Moseley (1913–1914) published his work on atomic numbers/ models, scientists had no idea or never asked the question as to whether there could be an underlying pattern (atomic theory) to explain periodicity. According to Robinson (2000), the gradual acceptance of the periodic table can be considered as a case of paradigm shifts in the Kuhnian sense, which can facilitate students' understanding: "Our students must come away with the notion that the chart as they know it (now arranged by atomic number and with the addition of quite a few elements) leaped from Mendeleeff's prediction to the inside cover of texts as if springing full-blown from the head of Zeus" (p. 177).

One general chemistry textbook presented the following description of the periodic table that ignored Mendeleev's fundamental assumption with respect to the atomic theory in the following terms:

> The periodic table was created by Mendeleev to summarize experimental observations. *He had no theory or model* to explain why all alkaline earths combine with oxygen in a 1:1 atom ratio — they just do (Moore et al. 2002, p. 266, emphasis added).

On the other hand, Moore (2003) has strongly endorsed the use of the history of the periodic table in the classroom:

> Asking students to argue pro or con for a particular representation of periodicity can be a challenging and instructive exercise. It requires that they know enough about properties of the elements to make convincing arguments, and it points out that science does not always arrive at a single, best, and correct answer to a complicated question (p. 847).

Interestingly, most textbooks and curricula try to convince students that science unequivocally provides 'single', 'best', and 'correct answers.'

A review of the literature in science education revealed that only one study (Ben-Zvi and Genut 1998) has been conducted to facilitate students' conceptual understanding of the periodic table based on a historical reconstruction of the numerous attempts to construct and improve its applications. These authors developed a teaching strategy that facilitated conceptual understanding of 8th and 10th grade Israeli students with respect to considering the periodic table as a scientific model rather than merely an inductive tool. Based on a history and philosophy of science perspective, these authors concluded: "... changing forms of the Periodic Table represent the heuristic power of the model on which these are based (Lakatos 1970). Had textbooks generally given some account of the changing form of the Table, it would encourage an approach more cognizant with scientific thinking" (Ben-Zvi and Genut 1998, p. 353). This brief review shows that a historical reconstruction of the development of the periodic table is essential for facilitating students' conceptual understanding of the periodic table. In this context, this study deals with the following research questions:

1. What was the role played by Dalton's atomic theory in the origin and development of the periodic table?
2. How could the role of periodicity as a function of the atomic theory be recognized before the modern atomic theory was postulated?
3. How could the discoverers of the periodic table predict the existence of new chemical elements and their properties?
4. How can we facilitate students' conceptual understanding of the periodic table by including various aspects related to the history and philosophy of science?

Based on these research questions and a review of the literature, the objectives of this study are:

1. Briefly review various aspects in the development of the periodic table based on a history and philosophy of science (HPS) perspective.
2. Design of a teaching strategy based on the historical aspects in order to facilitate freshman students' conceptual understanding of the periodic table.

Rationale of the Study

Almost every general chemistry textbook published on this planet recognizes the contribution of Mendeleev in the development of the periodic table. However, almost all these textbooks ignore that Mendeleev not only presented the periodic law to construct the periodic table but also:

(a) 'Speculated' with respect to the possible cause of the periodicity.
(b) Hypothesized with respect to the structure of the atom long before Thomson started his experiments in 1897.

Let us see how Mendeleev approached the subject in his own words at the famous Faraday Lecture delivered on June 4, 1889:

> The periodic law has clearly shown that the masses of the atoms increase abruptly, by steps, which are clearly connected in some way with Dalton's law of multiple proportions ... While connecting by new bonds the theory of the chemical elements with *Dalton's theory of multiple proportions, or atomic structure of bodies*, the periodic law opened for natural philosophy a new and wide field for speculation (Mendeleev 1889, p. 642, italics added).

In 1891 in his *Principles of Chemistry*, Mendeleev was even more thoughtful and at the same time perhaps more conscious of the history of science:

> To explain and express the periodic law is to explain and express the cause of the law of multiple proportions, of the difference of the elements, and the variation of their atomicity, and at the same time to understand what mass and gravitation are. In my opinion this is now premature. But just as without knowing the cause of gravitation, it is possible to make use of the law of gravity, so for the aims of chemistry it is possible to take advantage of the laws discovered by chemistry without being able to explain their causes (Reproduced in van Spronsen 1969, p. 61).

This clearly shows how Mendeleev was aware of the complexities involved in explaining the cause of the periodic law, and how this involved among other properties, atomicity. As a practicing chemist, he then assumes the role Newton chose in the case of the laws of gravity and motion (for further details see Conclusion section).

Based on a review of the history and philosophy of science (HPS) literature, among others, the following aspects in the development of the periodic table are particularly helpful in its conceptual understanding: (1) Role of the Karlsruhe Congress of 1860; (2) Accommodation of the chemical elements in the periodic table; (3) Prediction of elements that were discovered later; (4) Corrections of atomic weights; (5) Periodicity in the periodic table as a function of the atomic theory; and (6) Accommodation of argon in the periodic table.

Role of the Karlsruhe Congress of 1860

Most historians consider the Karlsruhe (Germany, September 3–5, 1860) Congress as crucial in the development of chemistry. The original idea of the congress was conceived by the German organic chemist A. Kekulé, in order to resolve differences among chemists with respect to the concepts of atom, molecule, atomic weight (atomic mass in modern terminology), valence, and among others. A circular (dated July 10, 1860) sent by the organizers of the congress to most outstanding chemists of Europe outlined its objective as the need to reach a consensus on, "More precise definitions of the concepts of atom, molecule, equivalent, atomicity, alkalinity, etc.; discussion on the true equivalents of bodies and their formulas; initiation of a plan for a rational nomenclature" (Reproduced in de Milt 1951, p. 421). Mendeleev, then 26 years of age and a postdoctoral researcher from St. Petersburg living in Heidelberg, attended the conference and was greatly impressed by S. Cannizaro's lecture. Interestingly, other researchers working on the periodic table also attended the congress (e.g., Meyer, Odling).

In a letter dated September 7, 1860, Mendeleev summarized the achievements of the congress:

> It is decided to take a different understanding of molecules and atoms, considering as a molecule the amount of a substance entering a reaction and determining physical properties, and considering as an atom the smallest amount of a substance included in a molecule. Further, it reached an understanding about equivalents, considered as empirical, not depending on the understanding about atoms and molecules (Reproduced in de Milt 1951, p. 422).

Indeed, deliberations at the Karlsruhe congress and Cannizaro's innovations had a lasting effect in the career of Mendeleev. One general chemistry textbook referred to the role played by the Karlsruhe Congress in the following terms:

In 1860, the Congress of Karlsruhe brought together many prominent chemists in an attempt to come to some agreement on issues such as the existence of atoms, the correct atomic masses, and how the elements are related to one another. No agreement was reached, but many attempts to explain new experimental data presented at the meeting were vigorously debated. One idea discussed was Avogadro's principle ... [which] allowed the relative atomic masses of the gases to be determined. Two scientists attending the congress were the German Lothar Meyer and the Russian Dmitri Mendeleev, both of whom left with copies of Avogadro's paper. In 1869, Meyer and Mendeleev discovered independently that a regular repeating of properties could be observed when the elements were arranged in order of increasing atomic mass (Atkins and Jones 2008, p. 38).

Accommodation of the Chemical Atoms in the Periodic Table

Availability of the atomic weights of about 60 elements in 1869 enabled Mendeleev to accommodate the elements in the table according to various physicochemical properties, such as: atomic weight, density, valence, specific heat, atomic volume, melting point, oxides, chlorides, and sulfides. Mendeleev (1879) enunciated his periodic law in cogent terms: "The properties of simple bodies, the constitution of their compounds, as well as the properties of these last, are periodic function of the atomic weights of elements" (p. 267). In contrast to other discoverers of the periodic table, Mendeleev's work was characterized by the division into main and sub-groups and the vacant spaces left for undiscovered elements. Elucidation of the concept of atomic weight and other properties by Cannizaro at Karlsruhe was crucial in the discovery of the periodic law by Mendeleev. Historically, Mendeleev's work has been referred to as a classification, system, table, or law and less frequently as a theory. According to Shapere (1977), the periodic table is neither a law nor a theory, but rather an ordered domain. Gordin (2004) has reconstructed Mendeleev's ideas from August 1869 to November 1870, to show how these evolved from considering his contribution a "regularity" to its "lawlike" character. As suggested by Erduran (2007), this lawlike character may not be in the same sense as the laws of physics, such as Newton's laws of motion. Giere (1999), however, goes beyond by suggesting a "science without laws." (For details see Niaz 2009a, Chap. 2). On the other hand, Weisberg (2007) has argued cogently that if theories allow us to unify, make predictions, and frame explanations, then Mendeleev must be considered as a theorist and his contribution a theory.

Brito et al. (2005) analyzed 57 general chemistry textbooks published in USA, and found that 55 presented a satisfactory description of the importance of accommodation in the periodic table. This shows that textbooks are well aware of the role played by accommodations. One textbook stated this in the following terms: "The arrangements of the elements in the periodic table correlate with the subshells that hold the highest-energy electrons" (Reger et al. 1997, p. 290). This was, of course, possible after the periodic law was enunciated in terms of the atomic number.

Prediction of Elements that were Discovered Later

Mendeleev left various vacant spaces in his table and made many predictions, of which the following are well-known: (a) Eka-aluminium (atomic weight = 68, density = 6.0, atomic volume = 11.5). This was discovered by the French chemist Paul Émile Lecoq de Boisbaudran in 1875, and was named gallium; (b) Eka-boron (atomic weight = 44, density = 3.5). This was discovered by the Swedish chemist Lars-Frederik Nilson in 1879, and was named scandium; (c) Eka-silicon (atomic weight = 72, density = 5.5, atomic volume = 13). This was discovered by the German chemist Clemens Alexander Winkler in 1886, and was named germanium. Actually, prediction of elements for the empty spaces was quite a complex process that involved various aspects of not only of the predicted element but also the surrounding elements and their properties in the periodic table. Philosopher of science, Wartofsky (1968) has explained this process in cogent terms:

> Mendeleev, for example, predicted that the blank space of atomic number 32 [Germa-nium], which lies between silicon and tin in the vertical column, would contain an element which was grayish-white, would be unaffected by acids and alkalis, and would give a white oxide when burned in air, and when he predicted also its atomic weight, atomic volume, density and boiling point, *he was using the periodic table as a hypothesis from which predictions could be deduced.* This was in 1871 (p. 203).

There is considerable controversy among historians and philosophers of science with respect to what made Mendeleev's law valid—accommodations dating from 1869 or the predictions from 1875 onwards. Despite the controversy, it is plausible to suggest that science educators can benefit from a consensus view, according to which both accommodations and predictions were important in the development of the periodic table (cf. Brush 2007; Niaz 2009a).

Brito et al. (2005) have reported that of the 57 general chemistry textbooks analyzed 30 explained the role of predictions satisfactorily. Interestingly, however, none of the textbooks satisfactorily explained the relative importance of both accommodations and predictions. Following presentation by one textbook came quite close to being satisfactory:

> Any good hypothesis must do two things: It must explain known facts, and it must make predictions about phenomena yet unknown … Mendeleev's hypothesis about how known chemical information could be organized passed all tests. Not only did the periodic table arrange data in a useful and consistent way to explain known facts about chemical reactivity, it also led to several remarkable predictions that were later found to be accurate (McMurry and Fay 2001, p. 160).

Corrections of the Atomic Weights

Besides the prediction of new elements, Mendeleev also corrected atomic weights of some of the existing elements. For example, atomic weight of beryllium was changed from 14 to 9, uranium changed from 120 to 240, and tellurium changed

from 128 to 125. One of the textbooks provided the following satisfactory presentation of this aspect:

> Two elements, tellurium (Te) and iodine (I), caused Mendeleev some problems.
> According to the best estimates at that time, the atomic mass of tellurium was greater than
> that of iodine. Yet if these elements were placed in the table according to their atomic
> masses, they would not fall into the proper groups required by their properties. Therefore,
> Mendeleev switched their order and in so doing violated his own periodic law. (Actually,
> he believed that the atomic mass of tellurium had been incorrectly measured, but this
> wasn't so (Brady et al. 2000, p. 63).

This shows the complexity of placing the elements in the periodic table and
how, in this case, although Mendeleev's decision was correct with respect to the
placing Te/I, but the atomic mass of Te is indeed greater than that of I.

Periodicity in the Periodic Table as a Function of the Atomic Theory

Many chemistry students must have wondered as to how Mendeleev and the other
co-discoverers could have conceptualized the underlying theoretical rationale of
the elements that manifested itself in periodicity. It is important to recall that most
of the pioneering work of Mendeleev and others was conducted from 1869 to
1889, before Thomson (1897), Rutherford (1911), Bohr (1913), and Moseley
(1913a, b) laid the foundations of the modern atomic theory. So how could
Mendeleev conceptualize periodicity as a function of the atomic theory? It is
important to note that given the positivist milieu of the scientific community in
which Mendeleev worked and lived, at times he did emphasize that the periodic
table was a legitimate induction from the verified facts. However, on other
occasions he was equally emphatic in recognizing the role of the atomic theory
(based on Dalton and subsequent changes, especially Cannizaro). Mendeleev's
dilemma was that, on the one hand, he could rightly claim that the periodic law
was based on experimental properties of the elements (following the positivist
milieu), and yet he could not give up the bigger challenge, namely the possible
causes of periodicity and hence importance of the atomic theory. Although in
1879, Mendeleev stated that he would not formulate an hypotheses (atomic theory), 10 years later in his famous Faraday Lecture, Mendeleev (1889) not only
attributed the success of the periodic table to Cannizaro's ideas on the atomic
theory (pp. 636–637) but went on to explicitly formulate the following hypothesis:

> ... the veil which conceals the true conception of mass, it nevertheless indicated that the
> explanation of that conception must be searched for in the masses of atoms; the more so,
> *as all masses are nothing but aggregations, or additions, of chemical atoms* ...(Mende-
> leev, 1889, p. 640, emphasis added).

For more details on this subject see Niaz et al. (2004). A comparison of the role played by hypotheses in the work of Newton and Mendeleev is presented in the conclusion section.

None of the general chemistry textbooks analyzed by Brito et al. (2005) presented this aspect satisfactorily. The closest that a textbook came to representing these issues was the following:

> Early in the nineteenth century, when Dalton's atomic theory was winning general acceptance, the first attempts were made toward classification of the elements into groups or families on the basis of similarities of physical and chemical properties ... even in its primitive form as stated in 1869, this [periodic] law clearly pointed to regularities that hinted at an orderly subatomic structure of matter and provided a tremendous stimulus toward seeking to understand the internal structure of atoms, as chemists and physicists sought to construct an atomic model that would explain Mendeleev's generalization (Sisler et al. 1980, p. 150).

Accommodation of Argon in the Periodic Table

Argon was discovered in 1895 and led to an intense debate with respect to its nature and place in the periodic table. According to Ramsay (1897):

> The discovery of argon at once raised the curiosity of Lord Rayleigh and myself as to its position in this table [Mendeleev's]. With a density of nearly 20, if a diatomic gas, like oxygen and nitrogen it would follow fluorine in the periodic table ... But when the ratio of its specific heats $[C_p/C_v]$ had, ... unmistakably shown that it was molecularly monoatomic, and not diatomic, as at first conjectured, it was necessary to believe that its atomic weight was 40, and not 20, and that it followed chlorine in the periodic table, and not fluorine. But here arises a difficulty. The atomic weight of chlorine is 35.5, and that of potassium, the next element in order in the table, is 39.1; and that of argon, 40, follows, and does not precede, that of potassium, as it might be expected to do (p. 379).

This shows how the placing of an element was not a straightforward question of ordering the elements in the ascending order of their atomic weights. Mendeleev himself took an active part in the controversy and considered the accommodation of argon as a glorious confirmation of the general applicability and validity of the periodic law. One of the textbooks refers to the discovery of argon in the following terms:

> In 1894, the British physicist Lord Rayleigh observed that a sample of nitrogen prepared by removing the oxygen, carbon dioxide, and water vapor from air had a density of 1.2572 grams per liter, whereas the liter of nitrogen prepared from ammonia had a density of only 1.2506 grams per liter under the same conditions. This difference caused Rayleigh to suspect a previously undiscovered element in the atmosphere. About the same time, Sir William Ramsay isolated a small amount of gas that would not combine with any other element by passing nitrogen obtained from the air over red-hot magnesium. Rayleigh and Ramsay both found that the residual gases showed spectral lines never before observed. In 1894, they announced the isolation of the first noble gas, which they called **argon**, meaning "the lazyone" (Holtzclaw and Robinson 1988, pp. 710–711).

Another textbook explicitly referred to the placement of argon in the periodic table:

> One problem with Mendeleev's table was that some elements seemed to be out of place. For example, when argon was isolated, it did not seem to have the correct mass for its location. Its relative atomic weight of 40 is the same as that of calcium, but argon is an inert gas and calcium a reactive metal. Such anomalies led scientists to question the use of relative atomic mass as the basis for organizing the elements (Atkins and Jones 2002, p. 42).

Interestingly, soon after pointing out the anomaly with respect to the placement of argon, and in the same paragraph, these authors go beyond by referring to a fundamental change in the underlying principle for classifying elements in the periodic table:

> When Henry Moseley examined x-ray spectra of the elements in the early twentieth century, he realized that all atoms of the same element had identical nuclear charge and, therefore, the same number of protons, which gives the element's atomic number. It was soon discovered that elements fall into the uniformly repeating pattern of the periodic table if they are organized according to atomic number, rather than atomic mass (Atkins and Jones 2002, p. 42).

In this *rationale* section, we have presented six aspects that were important for the development of the periodic table, and hence in its conceptual understanding. Besides presenting the historical and philosophical controversies, we have also included examples from the presentations of general chemistry textbooks that are used widely in various parts of the world. These examples show that at least some textbooks consider these aspects (generally considered to be part of history and philosophy of science) to be necessary for understanding the periodic table and chemistry (for further details, see Brito et al. 2005; Niaz et al. 2004). Furthermore, these details can help not only to understand results reported in this study but also in the design of future training studies.

Method

This study is based on two groups of freshman students enrolled in a Chemistry I course at a major university in Latin America. One group was randomly designated as the Control group and the other as the Experimental group. Both groups were taught by the same instructor (second author of this study). In order to avoid possible interactions between the two groups, control group students received instruction in a semester prior to that of the experimental group.

Control group ($n = 45$) students were asked to first consult the periodic table in the Internet and the following textbooks: Chang (2007) and Mahan and Myers (1990). There were other textbooks in the library and the students were free to consult them (some of these were analyzed by Brito et al. 2005). Next the instructor based on a traditional expository methodology presented the following aspects: early periodic systems (Döbereiner, Newlands, Meyer, and among others) Mendeleev's periodic table, classification and order of the elements, corrections in atomic weights, predictions of new elements, and contribution of Moseley (from atomic weights to atomic numbers). Most general chemistry textbooks provide considerable details about these aspects (cf. Brito et al. 2005). This phase of the study lasted about 2 weeks. A week later students were evaluated on Posttest 1, and later evaluated after about 2–3 weeks on Posttest 2. It is plausible to suggest that in this study a pretest was not necessary as students lacked basic information that was provided by the instructor, Internet and the textbooks, which constitutes the 'training' phase. Consequently, both Posttests 1 and 2 provide students' acquisition of various aspects of the periodic table, which facilitated conceptual understanding.

Experimental group ($n = 32$) students were also asked to consult the periodic table in the Internet and the following textbooks: Chang (2007) and Mahan and Myers (1990). Next, the instructor presented and discussed the same aspects of the periodic table that were used with the control group. A novel feature of this presentation was that it included considerations from the history and philosophy of science, such as (based on Brito et al. 2005): (a) Importance of accommodation of chemical elements according to their properties in the periodic table; (b) Importance of prediction of new elements as evidence for the periodic law; (c) Relative importance of accommodation and prediction in the development of the periodic table; (d) Illustrations of periodicity in the periodic table; (e) Contribution of Mendeleev: Theory or an empirical law?; and (f) Development of the periodic table as a progressive sequence of heuristic principles: Early ideas about atomic theory (e.g., Dalton) → Attempts to classify elements starting in 1817 → Mendeleev's first periodic table in 1869 → Discovery of argon in 1895 → Contribution of Moseley in 1913 based on atomic numbers. After this experience students were asked to construct concept maps (Novak 1990). Students had experience in the construction of concept maps as part of a course "Cognitive Development and Learning Strategies", in the previous semester. In the following week students were evaluated on Posttest 1. Based on students' responses to the four items of Posttest 1, classroom discussions helped to clarify different aspects of the periodic table. Next, in order to motivate students a PowerPoint presentation of various historical episodes was presented by the instructor. Some students requested a copy of the PowerPoint presentation, which was provided to them. After this, students were invited to construct concept maps again, which were discussed in class and compared to their previous concept maps. All these activities lasted about 3–4 weeks, at the end of which students were evaluated on Posttest 2. Finally, five volunteer students participated in semi-structured interviews, each of which lasted about 45 min. Both the control and experimental groups were tested on Posttests 1 and 2.

Posttest 1

Item 1: In your opinion what was the criterion used by Mendeleev to put the elements in the established order in the periodic table?

Item 2: If the periodic table was elaborated before the modern atomic theory, do you think there is a relationship between the periodic table and the earlier atomic theory?

Item 3: If the periodic table was elaborated before the modern atomic theory, how could Mendeleev and others construct the periodic table?

Item 4: Did the idea of ordering the elements originate with Mendeleev's periodic table?

Posttest 2

Item 5: In your opinion, in the acceptance of the periodic table, which of the following factors was most important?

(a) Accommodation of the chemical elements that is classification according to their physicochemical properties.
(b) Prediction of some of the elements that were discovered later.
(c) Corrections of the atomic weights of some of the elements.
(d) No/ambiguous response.

Item 6: In your opinion, which factors were important for the development of the periodic table? (*Note* In this item students generated their own factors, which are presented in the Results and Discussion section).

Item 7: Periodicity of elements in the periodic table is: A consequence of physically observable properties (as aggregates) or chemical atoms as particles?

Validation of Students' Responses on Items in Posttests 1 and 2

Responses of five students from each of the two groups (control and experimental) were selected randomly. This constituted a pool of 70 responses (each of the 10 students selected had seven responses), that were classified by both authors according to the criteria for each item (Items are presented in this section and also in the Results and Discussion section). Most of the criteria for classification were the same as used by Niaz et al. (2002), namely conceptual and rhetorical. In general, a conceptual response showed an understanding of the underlying issues, whereas a rhetorical response simply reiterated the information provided. Examples of both types of responses are provided in the next section for both the control and experimental groups. There were disagreements in the classification of 11 (16 %) responses. These were discussed in various meetings and both authors presented arguments and finally a consensus was achieved. Remaining responses were then classified by the second author.

Results and Discussion

Criterion Used by Mendeleev to put the Elements in the Established Order (Students' Responses on Item 1, Posttest 1)

Responses of both groups of students on Item 1 of the Posttest 1 are presented in Table 1.

Table 1 shows that the difference in the performance of experimental (19 %) and control (2 %) group students who gave a conceptual response is statistically significant ($p < 0.05$). Following is an example of a conceptual response by an experimental group student:

> In order to construct the periodic table *Mendeleev took the atomic theory as his base*. At the same time he also used the periodic law to refute some of the experimental results, as he used the physical properties of the elements, such as density, atomic weight, valence and the oxides (Student #10, italics added).

At this stage, it is interesting to compare the concept map constructed by Student #10 (see Fig. 1), and his response on Item 1. In his response, the student considers the atomic theory as Mendeleev's based and at the same time recognizes the importance of properties such as: density, atomic weight, valence, and the oxides. Now let us have a closer look at the concept map in which the following linkages (among others) play an important role in the development of the periodic table: (a) Periodic table and Mendeleev 1869; (b) Periodic table and Dalton; (c) Dalton and atomic theory; (d) Atomic theory and (atomic weight, valence, equivalent weight, and density); (e) Physical and chemical properties and (atomic weight, valence, equivalent weight, and density); (f) Mendeleev 1869 and physical and chemical properties; and (g) Avogadro clarified the doubts between Gay-Lussac and Dalton. Interestingly, the concept map was constructed about a week before the student responded to Item 1 of the Posttest 1. It is plausible to suggest that the linkages included in the concept map were fairly comprehensive and representative of the classroom treatment for the experimental group. Furthermore, while responding to Item 1, the student clearly recalled these linkages and then used them creatively to facilitate conceptual understanding. Perhaps, most important of all was the recognition by this student of the role played by

Table 1 Comparison of the performance of control and experimental group students on item 1[a] (Posttest 1)

Response	Control ($n = 45$)	Experimental ($n = 32$)	χ^2 (Sig.)
Conceptual	1 (2 %)	6 (19 %)	4.34 ($p < 0.05$)
Rhetorical	34 (76 %)	22 (69 %)	ns
No response	10 (22 %)	4 (13 %)	ns

[a] Item 1: In your opinion what was the criterion used by Mendeleev to put the elements in the established order in the periodic table?

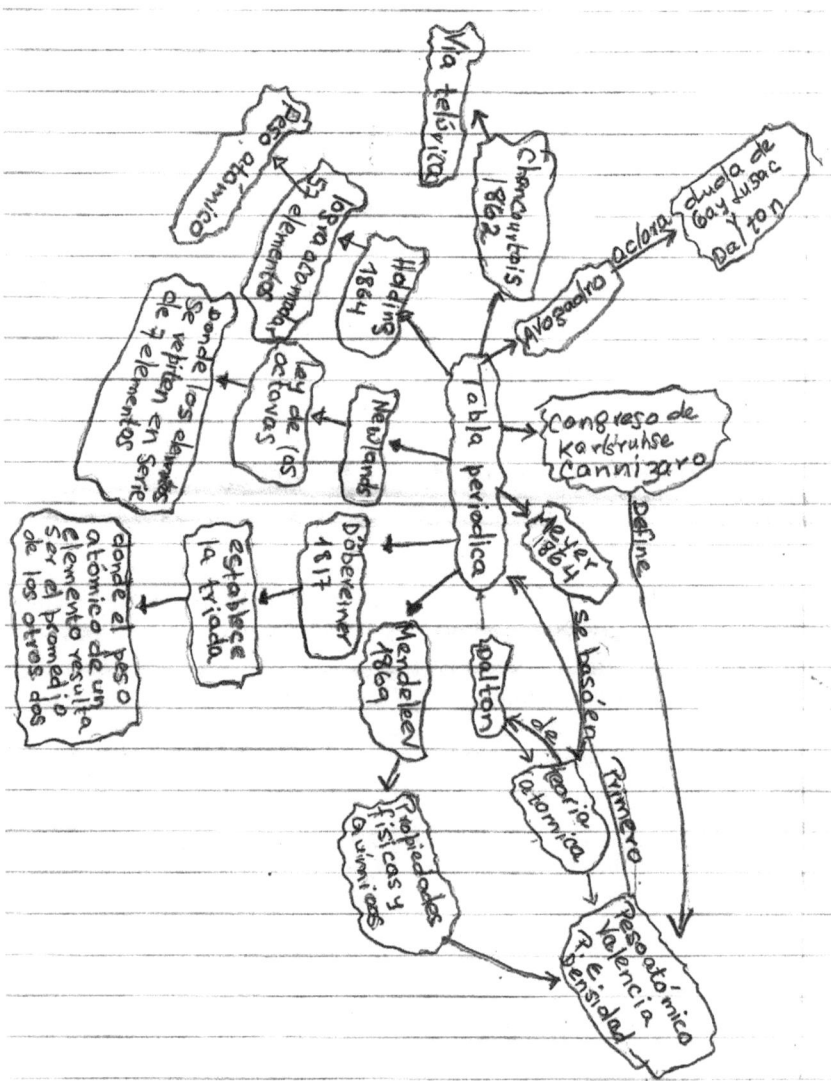

Fig. 1 Concept map drawn by Student #10. *Note* This concept map is reproduced from the student's notebook. In order to facilitate visibility all the words are retraced with pencil #2. There are some mistakes with respect to some names and properties, which were part of the original concept map

Avogadro in the development of both the atomic theory and the periodic table. This shows that the concept maps were helpful in facilitating conceptual understanding by integrating various aspects discussed in class.

Following is an example of a conceptual response by a control group student:

> *Mendeleev employed a method according to which every material must be composed of molecules or smaller particles.* Then he started to study the compounds and discovered that these have elements with a particular weight, which was then used to order the elements in the periodic table. Consequently, the atomic weight became the criterion for ordering the elements (Student #9, italics added).

It is interesting to note that even one student from the control group understood fairly well Mendeleev's underlying hypothesis, namely the periodic nature of the physical and chemical properties of the elements is a consequence of the atomic nature of all matter. How do we explain a conceptual response from a control group student who did not have the benefit of the experimental treatment? This is an important issue and will be discussed in the conclusion section.

According to Table 1, a majority of the control (76 %) and experimental (69 %) group students gave a rhetorical response, which reiterated that the criterion used by Mendeleev to order the elements was the atomic weight. Of course, atomic weights could be determined experimentally in Mendeleev's time, whereas atomic nature of matter was still to be demonstrated experimentally. The inductivist milieu prevalent in the late nineteenth century reinforced the role played by atomic weights and this continues to be the case even in our present day chemistry textbooks (cf. Brito et al. 2005). Interestingly, Mendeleev was fully aware of this dilemma ("The veil which conceals the true conception of mass", cf. Mendeleev 1889, p. 640) and in his writings clearly expressed the need to recognize the atomic nature of matter as the underlying cause of the periodicity of properties of elements (cf. Rationale section). Following is an example of a rhetorical response by an experimental group student:

> Mendeleev took extreme care while explaining his periodic table. In 1869 he presented a study that dealt with the *nature of the chemical elements as a function of their atomic weights*. This helped him to order the elements in the periodic table (Student #17).

Following is an example of a rhetorical response by a control group student:

> Mendeleev based some of his work on Meyer's attempt *to order the elements according to their atomic weight.* Mendeleev suggested that elements having different properties could not be placed in the same group, such as the metals and the non-metals. Furthermore, he ordered the elements in the increasing order of atomic weights (Student #4).

It is interesting to compare the conceptual and rhetorical responses of both the experimental and control group students. Conceptual responses specifically refer to the particulate nature of matter (atoms or molecules), whereas the rhetorical responses refer to the atomic weights as a criterion for the classification of elements (see the parts in italics). In our opinion, rhetorical responses are not necessarily wrong, but rather lack the additional insight with respect to the underlying reason for periodicity of elements in the periodic table.

In order to further appreciate the difference between a rhetorical and a conceptual response, let us consider the following responses by two experimental group students:

Mendeleev (1869) discovers his periodic system, *employs chemical and physical properties*, also employs atomic weights in order to support his system, and enunciates that elements when placed according to their atomic weights present a clear periodicity in their properties (Student #28).

Mendeleev studied the properties of the elements, such as atomic weight, *oxides formed by the reaction with oxygen, density, volume — that is chemical and physical properties*. Besides this on occasions he used to alter the atomic weights of some elements in order to make them concordant with the established order. Such criteria permitted him to place new elements in his periodic table, such as eka boron, eka aluminium and eka silicon (Student #4).

Response by Student #28 was classified as rhetorical, whereas the one by Student #4 was classified as conceptual. It is important to note that both responses can be considered as correct. Nevertheless, the conceptual response provides additional information with respect to: (a) Provides examples of physical and chemical properties (whereas the rhetorical response simply refers to the properties, see the part in italics in both responses); (b) Even after having enunciated his law, Mendeleev used to alter the atomic weights of some elements and hence their place in the periodic table; (c) Mendeleev's table not only classified the known elements but also predicted new ones. It is plausible to suggest that the conceptual response is a better indicator of how scientists when faced with difficulties and anomalous data, accept changes in their enunciated laws, which are basically working hypotheses and hence tentative.

Relationship Between the Periodic Table and the Earlier Atomic Theory (Students' responses on Item 2, Posttest 1)

Responses of both groups of students on Item 2 of the Posttest 1 are presented in Table 2.

Table 2 shows that the difference in the performance of experimental (47 %) and control (4 %) group students who gave a conceptual response is statistically significant ($p < 0.001$). Following is an example of a conceptual response by an experimental group student:

Table 2 Comparison of the performance of control and experimental group students on item 2[a] (Posttest 1)

Response	Control ($n = 45$)	Experimental ($n = 32$)	χ^2 (Sig.)
Conceptual	2 (4 %)	15 (47 %)	17.18 ($p < 0.001$)
Rhetorical	25 (56 %)	4 (13 %)	12.99 ($p < 0.001$)
No response	18 (40 %)	13 (41 %)	ns

[a] Item 2: If the periodic table was elaborated before the modern atomic theory, do you think there is a relationship between the periodic table and the earlier atomic theory?

There is a relationship between the atomic theory and the periodic table, because the early periodic systems of the elements were fundamentally based on the atomic weights. Atomic weight itself was proposed by Dalton and with it started the atomic theory (Student #9, italics added).

Following is an example of a conceptual response by a control group student:

There is a relationship between the two. Although there was no definitive study of the atomic theory, some of the properties of the atoms and matter were known that facilitated the discovery of the periodic table based on the ordering of the elements according to their physical and chemical properties. The relationship became obvious quite early as the chemists tried to study the properties of the elements and later attempts at ordering them. These *early chemists must have had some notions of the atomic theory in order to start exploring the periodic table*. It is precisely these initial theories of the atom that constituted the base of the modern atomic theory (Student #33, italics added).

These and other examples are good evidence as to how if students are provided an opportunity to think, explore, and imagine, this can lead to fairly plausible scenarios of how the atomic theory led to the genesis of the periodic table (see Conclusion section for details, and also italics in both responses).

Table 2 shows that 56 % of the control group students and 13 % of the experimental group students gave a rhetorical response and the difference was statistically significant ($p < 0.001$). This clearly shows that the experimental group students understood the problem situation much better as compared to Item 1, and consequently the rhetorical responses decreased from 69 % (Item 1) to 13 % on Item 2. Following is an example of a rhetorical response by an experimental group student:

Although the periodic table was elaborated long before the atomic theory there is a relationship between the two. Elaboration of the periodic table is based principally on the atomic weights of the elements, which were in turn based on the periodic table (Student #7, italics added).

This response was classified as 'rhetorical' as it does not seem to differentiate between the atomic theory and the modern atomic theory (that is after 1897, based on experiments by J. J. Thomson and others). It appears that the student seems to be suggesting that the periodic table anteceded the atomic theory as proposed by Dalton and others (see the part in italics). Furthermore, the student suggests that the atomic weights of the elements were based on the periodic table, instead of stating that the atomic weights were based on the atomic theory and in particular that of Dalton. Some evidence for this line of reasoning can also be observed in the concept map drawn by this student (See Fig. 2, Concept map of Student #7).

First, it is important to note that the concept map drawn by Student #7 is a fairly good representation of the development of the periodic table. It highlights all the important linkages between the periodic table and its important discoverers. Furthermore, it clearly establishes a linkage between Dalton's atomic theory and Meyer's periodic table. However, in the case of Mendeleev, although it does refer to the atoms and Dalton's atomic theory, it does not explicitly establish a linkage between Dalton's atomic theory and Mendeleev's periodic table. It is possible that

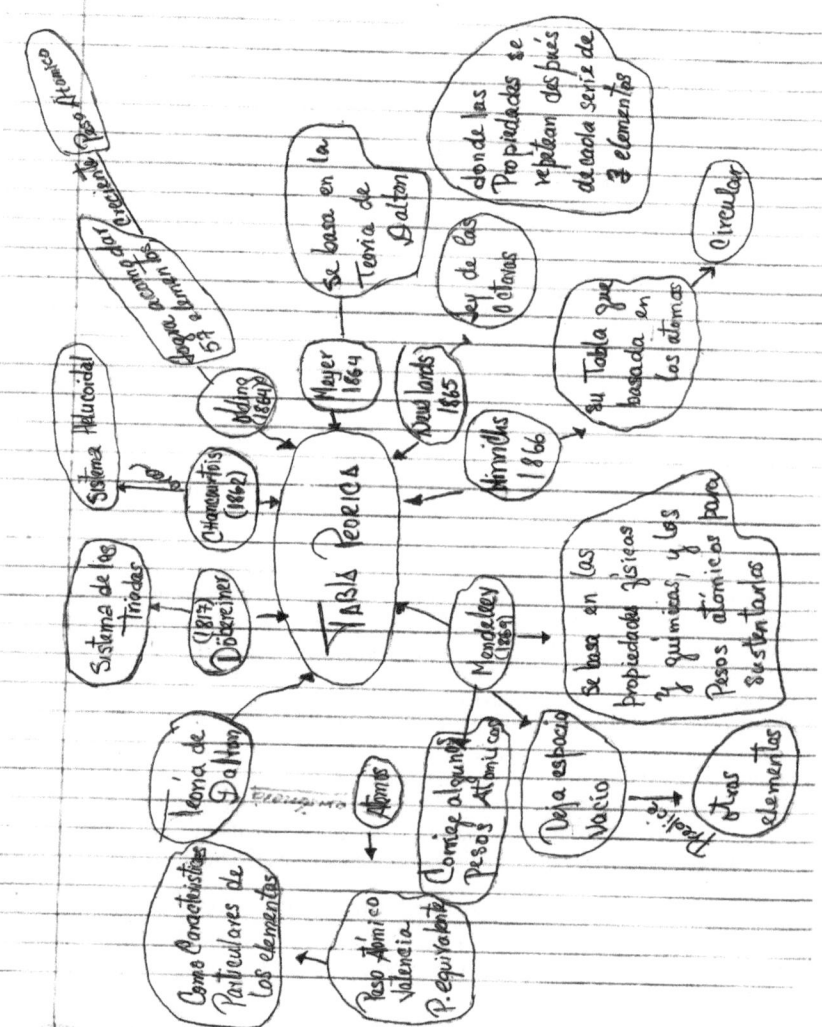

Fig. 2 Concept map drawn by Student #7. *Note* This concept map is reproduced from the student's notebook. In order to facilitate visibility all the words are retraced with pencil #2. There are some mistakes with respect to some names and properties, which were part of the original concept map

this ambiguity in the understanding of Student #7 is reflected in his/her response on Item 2. Leaving aside this difficulty in student's understanding, this concept map does establish a linkage between Mendeleev's contribution and the following: correction of atomic weights, empty spaces in the periodic table, prediction of new elements, and that the table is based on physicochemical properties and the atomic weights.

Following is an example of a rhetorical response by a control group student:

> The periodic table is based on the atomic theory, in which the elements are classified according to their atomic numbers. *If the periodic table had been elaborated before the atomic theory*, the classification of the elements would have been very different, based on criteria such as the physical and chemical properties (Student #1, italics added).

This response was classified as 'rhetorical' as it does not differentiate between the modern atomic theory and that proposed by Dalton and others before the first attempts to elaborate the periodic table (see the part in italics). Furthermore, it explicitly denies the possibility of a relationship between the early atomic theory and the periodic table. It is important to note that almost all general chemistry textbooks, both at the high school and university level courses clearly establish the difference between the classical atomic theory (early and middle nineteenth century) and the modern atomic theory (starting in the late nineteenth and early twentieth century) and following is an example:

> Dalton formulated his atomic theory which can be summarized in five short statements [not reproduced here as these are well known] ... Why do atoms of one element combine with only certain numbers of atoms of other elements in forming chemical compounds? Dalton wanted a physical model — a picture of the atom — that fits his theory and helped explain it. Dalton claimed that if you accepted the five statements of his theory as a viable model of reality, the laws of conservation of matter and constant composition could be easily explained and accounted for. Many scientists at the time did not accept Dalton's theory or his model and continued to argue that atoms did not exist ... Nearly 200 years after Dalton proposed his atomic model, we now know that none of the five statements of his theory are *entirely* true. Atoms are not the most fundamental of particles—they are composed of even smaller particles we call electrons, protons, and neutrons. None of these later findings should be taken as diminishing Dalton's accomplishments, however. For the time, his was a superb model ... It took nearly 100 years for the last of the holdouts to become convinced (or to die off!), but by the beginning of the twentieth century the concept of the atom was firmly established (Russo and Silver 2002, pp. 78–80).

Similar statements in most general chemistry textbooks provide students with a historical perspective in which Dalton's theory originated, explained some experimental findings, critiqued, and then replaced by the modern atomic theory in the early twentieth century.

In order to further appreciate the difference between rhetorical and conceptual responses, let us consider the following responses by two experimental group students:

> Yes [relationship between the periodic table and the earlier atomic theory, Item 2], as this helped to explain why some elements were not concordant with the table proposed by Mendeleev, as these appeared to be out of place (Student #30).

> The periodic table did not emerge before the atomic theory. On the contrary, the study of the elements first started with the help of Dalton's atomic theory, and years later the first periodic tables appeared. Yes there is a relationship, as *in the periodic table the elements were ordered according to their atomic weights, and these in turn were based on the atomic theory* (Student #23, italics added).

In some sense, both responses (Student #23 and #30) can be considered as correct. However, it seems that the response by Student #30, does not deal explicitly with the genesis of the periodic table and hence was classified as rhetorical. On the other hand, the response by Student #23 was classified as conceptual, due to the following aspects: (a) Clearly states that the elaboration of the periodic table depended on Dalton's atomic theory; (b) Elements in the early periodic tables were ordered according to their atomic weights; and (c) Atomic weights in turn required the conceptualization of the atomic theory (see the part in italics).

Relationship Between the Periodic Table and the Modern Atomic Theory (Students' Responses on Item 3, Posttest 1)

Responses of both groups of students on Item 3 of the Posttest 1 are presented in Table 3.

Table 3 shows that the difference in the performance of experimental and control group students who gave a conceptual response is statistically significant ($p < 0.01$). Following is an example of a conceptual response by an experimental group student:

> In the year 1817, Döbereiner proposed a system of triads based on three elements, in which the atomic weight of one of the elements was a mean of the other two. In 1860 at the Karlsruhe Congress, the concepts of atomic weight, equivalent weight and the atom were clarified. In 1866, Hinrichs proposed a periodic system based on the shape of the elements. Later Meyer presented his periodic table based on the atomic theory. *In 1869 Mendeleev proposed a periodic table based on the physical and chemical properties and the atomic theory.* Mendeleev's periodic system was ingenious as he left empty spaces in his table for predictions of new elements and corrections. For example, a monoatomic gas [argon] was discovered that questioned the very existence of the periodic table, as it was difficult to find a place for it. Finally, it was proposed that it could be placed in a separate group between the halogens and alkaline metals. Forty years later, the electron was discovered and it was decided that the position of an element in the periodic table was determined by the number of electrons (Student #9, emphasis added).

This is a very interesting and satisfactory response as it posits Mendeleev's contribution within a historical perspective. In other words, Mendeleev was not alone in the quest for understanding the classification of the chemical elements.

Table 3 Comparison of the performance of control and experimental group students on item 3[a] (Posttest 1)

Response	Control ($n = 45$)	Experimental ($n = 32$)	χ^2 (Sig.)
Conceptual	1 (2 %)	9 (28 %)	8.92 ($p < 0.01$)
Rhetorical	20 (44 %)	12 (38 %)	ns
No response	24 (53 %)	11 (34 %)	ns

[a] Item 3: If the periodic table was elaborated before the modern atomic theory, how could Mendeleev and others construct the periodic table?

There were many 'players' that helped to provide a better understanding, and this student (also some others) has highlighted the role of the following: (a) Döbereiner's triads; (b) Karlsruhe congress that was attended among others by Mendeleev, Meyer and Cannizaro. Elucidation of the concept of atomic weight by Cannizaro at the Congress was particularly helpful to both Mendeleev and Meyer; (c) Hinrichs' contribution; (d) Meyer's contribution; and (e) Place of argon in the periodic table. Finally, and most important of all this student has referred to the role played by the atomic theory (see the part with emphasis in the response). Brito et al. (2005) have recommended the use of such a sequence of heuristic principles in order to facilitate students' understanding (see Rationale section).

One student from the control group provided the following conceptual response:

> This becomes quite clear if we study the contributions of the early chemists who tried to study the properties of the elements and later classify them. These persons must have had *some notions of the atomic theory* in order to start their studies about the periodic table. The early atomic theory that was used by these chemists can be considered as the base of the modern atomic theory (Student #33, emphasis added).

Once again this is a good example of a conceptual response as it refers to one of the most important antecedent (that is atomic theory) of the early work on the periodic table that was particularly helpful to Mendeleev and others. It is interesting to note that this student (#33) provided a very similar conceptual response to Item 2, which shows some consistency in thinking. Following is an example of a 'rhetorical' response by a student from the experimental group:

> *Before the atoms appeared chemists were already studying the elements of the periodic table.* They started to study the elements even when all of them were not classified. Through the study of the elements, chemists started to study the atomic theory (Student # 16, italics added).

This response was classified as 'rhetorical' as it first asserts that the study of the elements started before the chemists knew about the atoms, and later adds that the elements were studied through the atomic theory (see the part in italics). This is not necessarily a wrong response but rather emphasizes aspects that may have occurred differently and hence rhetorical. A history and philosophy of science perspective shows that it was the atomic theory which provided the chemists the underlying rationale for classifying the elements. Following is an example of a 'rhetorical' response by a student from the control group:

> It started with Döbereiner who proposed his law of the triads, and later we had Newlands with his law of the octaves, which could not explain correctly the order of the elements. To go beyond the work of Newlands, *Mendeleev accommodated the periodic table based on atomic weights*. Later Meyer observed that there was relationship between the atomic weight and the volume of an atom. Finally, Moseley explained that Mendeleev's doubts with respect to the placement of the elements could be resolved if the elements were ordered according to their atomic numbers (Student #18, italics added).

This is a fairly good response based on a historical reconstruction and quite similar to the one provided by Student #9 on this item (see above). However, it was classified as 'rhetorical' as it does not explicitly distinguish with respect to how Mendeleev and others could elaborate the periodic table without the modern atomic theory (see the part in italics).

In order to further illustrate the difference between rhetorical and conceptual responses, let us consider the following responses by experimental group students:

> Mendeleev and others were able *to construct the periodic table through the atomic weights of the elements.* Besides they also used other data and after many years were able to present what we today call the periodic table (Student # 22, italics added).

> First they studied the compounds, molecules and everything related to the elements. Later, many chemists started to contribute towards the development of the periodic table and each one of them corrected and constructed a new table. Some of these scientists were the following: Döbereiner (triads), Chancourtois (helical system), Odling, Meyer (periodic law), *Cannizzaro (Karlsuhe Congress, atomic weights, equivalent weights, valence)*, Mendeleev and others. Each one of them contributed by constructing a periodic table based on the knowledge that was available (Student #2, italics added).

Most readers would agree that both responses are basically correct. However, response of Student #22 was classified as rhetorical as it simply states the importance of atomic weights in the development of the periodic table (see the part in italics). On the contrary, response of Student #2 was classified as conceptual as it explicitly deals with the following aspects: (a) Role played by the elements, molecules and the formation of compounds; (b) Development of the periodic table as a collective enterprise; and (c) Role of Cannizaro at the Karlsruhe Congress and elucidation of various chemical concepts (see the part in italics).

Ordering of the Elements and Mendeleev's Periodic Table (Students' Responses on Item 4, Posttest 1)

Responses of both groups of students on Item 4 of the Posttest 1 are presented in Table 4.

The idea behind this item was to explore students' understanding of the development of the periodic table within a research tradition that was of interest to the scientific community. It was the collective efforts of this community of which Mendeleev was a member, which took many years and tentative attempts to understand the classification of the chemical elements. In other words, Mendeleev

Table 4 Comparison of the performance of control and experimental group students on item 4[a] (Posttest 1)

Response	Control ($n = 45$)	Experimental ($n = 32$)	χ^2 (Sig.)
Conceptual	–	4 (13 %)	–
Rhetorical	37 (82 %)	18 (56 %)	4.97 ($p < 0.05$)
No response	8 (18 %)	10 (31 %)	ns

[a] Item 4: Did the idea of ordering the elements originate with Mendeleev's periodic table?

was continually critiqued and also received feedback from his contemporaries. Table 4 shows that none of the students from the control group and four (13 %) students from the experimental group provided a conceptual response. Following is an example of a conceptual response from an experimental group student:

> It originated with Döbereiner and his triads, which had to be discarded as many elements could not be included in the ordering. Before this there were also the contributions of Dalton and Gay Lussac that helped in the formulation of the periodic tables. De Chancourtois formulated a periodic table based on helical graphical system. Later Meyer proposed his periodic table. Mendeleev corrected the atomic weights and predicted new elements in his periodic table (Student #2).

This presentation explicitly refers to the work of others who anteceded Mendeleev with respect to the ordering of elements in the periodic table. Following is an example of a rhetorical response by a student from the control group:

> Mendeleev was the first to have the idea of ordering the elements and did so for the elements 1-88. Although he ordered the elements in ascending order of their atomic weights, he had doubts with respect to the place of some of the elements. It was these difficulties faced by Mendeleev that required the help of other chemists, especially Meyer and Moseley. Moseley ordered the elements according to the ascending order of their atomic numbers (Student #18).

Following is an example of a rhetorical response by a student from the experimental group:

> Yes, *the idea of ordering appeared first in the periodic table of Mendeleev* in the middle of the 19th century. However, by 1913-14, the table was completely elaborated, based on the empty spaces left by Mendeleev, which were filled by the work of other scientists (Student #24, italics added).

These two responses are good examples of the ahistoric understanding of the periodic table in which almost all credit is given to Mendeleev and the role of a critical community is ignored (see the part in italics). It is surprising that even some of the experimental group students expressed such views, as these students were exposed to a historical reconstruction at various opportunities during the semester. In part, this reflects the influence of ahistoric presentations of some textbooks.

Factors that were Important in the Acceptance of the Periodic Table (Students' Responses on Item 5, Posttest 2)

Responses of both groups of students on Item 5 of the Posttest 2 are presented in Table 5.

Table 5 Comparison of the performance of control and experimental group students on item 5[a] (Posttest 2)

Response	Control ($n = 45$)	Experimental ($n = 32$)	χ^2 (Sig.)
a	21 (47 %)	11 (34 %)	ns
b	6 (13 %)	1 (3 %)	ns
c	1 (2 %)	9 (28 %)	8.93 ($p < 0.01$)
d	11 (24 %)	2 (6 %)	ns
e	4 (9 %)	4 (13 %)	ns
f	2 (4 %)	5 (16 %)	ns

[a] Item 5: In your opinion, in the acceptance of the periodic table, which of the following factors was most important?
(a) Accommodation of the chemical elements that is classification according to their physico–chemical properties
(b) Prediction of some of the elements that were discovered later
(c) Corrections of the atomic weights of some of the elements
(d) No/ambiguous response
(e) Combination of two factors (a, b, c or d)
(f) Combination of three factors (a, b, c or d)

Examples of Students' Responses who Selected Option (a)

The idea behind this item was to provide students with various options, all of which were relevant and important in the development of the periodic table and have been recognized as such, both in the history and philosophy of science (Brush 1996) and science education literature (Brito et al. 2005). Option (a), accommodation of the chemical elements was selected by 21 (47 %) students from the control group and 11 (34 %) students from the experimental group and the difference between the two is statistically not significant. Brito et al. (2005) reported that almost all the textbooks ($n = 57$, published in USA) in their study, satisfactorily explained the role of accommodations in the development of the periodic table. As both the control and experimental groups consulted at least some of these textbooks, it is understandable that at least one-third of the students in both groups accepted this factor. Following is an example of a response by one of the students from the control group:

> From the beginning the elements were organized in the periodic tables according to their atomic weights, as it was believed that the properties would be related to it. With this organization most of the elements that were in the same group coincided in their physicochemical properties. However, Mendeleev found spaces in which he could not place the elements even when they had similar properties as the atomic weight did not coincide with the ascending order required for organization. This led to the idea that the elements could not be organized according to their atomic weight but rather the atomic number (Student #1).

This is an interesting response as it refers not only to the physicochemical properties as a factor but also refers to the difficulties involved with the organization of the elements. Following is an example of a response by one of the students from the experimental group:

The most important factor was the accommodation of the elements according to their physicochemical properties. Mendeleev accommodated the elements according to their atomic weight, valence and equivalent weight, and took extreme care to avoid errors. However, when he had almost finished the elaboration of the periodic table, there appeared an inert gas that questioned the basis of the periodic table. Finally, this gas [Argon] was placed in a new group between the halogens and the alkaline metals, which helped to accommodate this element that was found to be inert (Student #3).

Examples of Students' Responses who Selected Option (b)

Option (b) was selected by only 6 (13 %) students from the control group and 1 (3 %) from the experimental group. Although textbooks do not assign the same importance to predictions as to accommodations (Brito et al. 2005), low percentage of students' responses in this study are somewhat unexpected. Especially, experimental group students did have the opportunity to discuss the importance of predictions. Following is an example of a response by one of the students from the experimental group:

The prediction of some of the elements that were later discovered is very important. These were based on a study of the atomic weights (Student #5).

Following is an example of a response by one of the students from the control group:

Ordering of elements according to their increasing or decreasing physicochemical properties led Mendeleev to the discovery of new elements. Newlands periodic table did not provide this information (Student #45).

It is interesting to note that none of the students from the two groups who selected this option provided examples of the elements that were predicted by Mendeleev and others.

Examples of Students' Responses who Selected Option (c)

Only 1 (2 %) student from the control group and 9 (28 %) from the experimental group selected option (c), and the difference in performance is statistically significant ($p < 0.01$). It is well-known that besides other corrections in the atomic weights of the elements, Mendeleev explicitly corrected the atomic weight of beryllium (9 instead 14), uranium (240 instead of 120), and tellurium (125 instead of 128). These corrections facilitated the accommodation of these elements according to their physicochemical properties. It is plausible to suggest that at least some of the experimental group students considered these changes as corrections instead of predictions. This perhaps explains why only one student from the experimental group accepted option (b) that is the importance of predictions. Interestingly, Brush (1996) considers the corrections in the atomic weights of Be, U, and Te as novel predictions. Following is an example of a response by one of the students from the experimental group:

Given the many errors in the determination of the atomic weights, Mendeleev had considerable difficulty in the classification of various elements. Based on various properties of the elements such as density, valence and others, Mendeleev suggested corrections in the atomic weights and thus changed the position of the elements (Student #31).

Following is an example of a response by one of the students from the control group:

Although there was fair amount of consensus among scientists with respect to atomic weight as the defining criterion for classification of the elements, a major problem was the difficulties involved in the classification of various elements that did not follow the relation between the ascending order of atomic weights and the various physicochemical properties. In the case of the three pairs of elements, the corrections introduced by Mendeleev were particularly helpful in overcoming the difficulties faced by the periodic table (Student #33).

Interestingly, the responses by these two students are quite similar, which show the influence of the textbooks, in which the corrections of atomic weights are generally discussed (cf. Brito et al. 2005).

Examples of Students' Responses who Selected Option (e) or Option (f)

Interestingly, some of the students in both groups preferred to combine different factors in order to respond to this item. In the experimental group, 13 % of the students selected option (e) by combining accommodation and correction of atomic weights, whereas in the control group 9 % of the students selected option (e) by combing accommodation and prediction. Similarly, in the experimental group 16 % of the students selected option (f) and in the control group 4 % of the students selected this option, which was based on the combination of three factors, namely accommodation, prediction, and correction of atomic weights.

Finally, it seems that of the three factors that were provided to the students, experimental group students seem to prefer accommodation and correction of atomic weights. On the other hand, control group students seem to prefer accommodation and prediction. It is plausible to suggest that this difference can be attributed to the influence of the experimental teaching strategy for the experimental group and textbook presentations (cf. Brito et al. 2005) for the control group.

Factors that were Important in the Development of the Periodic Table (Students' Responses on Item 6, Posttest 2)

Responses of both groups of students on Item 6 of the Posttest 2 are presented in Table 6.

Table 6 Comparison of the performance of control and experimental group students on item 6[a] (Posttest 2)

Response	Control ($n = 45$)	Experimental ($n = 32$)	χ^2 (Sig.)
a	29 (64 %)	9 (28 %)	8.47 ($p < 0.01$)
b	–	5 (16 %)	–
c	–	4 (13 %)	–
d	–	2 (6 %)	–
e	16 (36 %)	12 (38 %)	ns

[a] Item 6: In your opinion, which factors were important for the development of the periodic table? (*Note* In this item students generated their own factors, which are presented below)
(a) Determination of the atomic weights, Döbereiner's triads, Newland's octaves, Mendeleev's periodic table, Meyer's periodic table, predictions of the properties of elements not discovered, corrections of atomic weights, Moseley's discovery of atomic number
(b) Dalton's atomic theory
(c) Karlsruhe Congress
(d) Placement of the noble gases
(e) No response

The idea behind this item was to let students suggest the factors that in their opinion were important for the development of the periodic table. In this sense this item is different from Item 5 in which the students were given a series of factors and they selected one or a combination of different factors. A comparison of student performance on Items 5 and 6 could provide feedback with respect to how students understand the development of the periodic table and the degree to which it is influenced by textbooks (control group) and the classroom treatment (experimental group). In the classification of students' responses on this item, following additional criteria were used: (i) If the student referred to all or some of the factors included in response (a) it was categorized as option (a); (ii) If the student referred to some of the factors included in response (a) and at the same time explicitly referred to Dalton's atomic theory, then it was categorized as option (b); (iii) If the student referred to some of the factors included in response (a) and at the same time explicitly referred to the Karlsruhe congress, then it was categorized as option (c); (iv) If the student referred to some of the factors included in response (a) and at the same time explicitly referred to the placement of noble gases, then it was categorized as option (d).

Examples of Students' Responses who Suggested Option (a)

A majority (64 %) of the control group and 28 % of the experimental group students referred to some or all of the factors included in option (a), and the difference in performance of the two groups is statistically significant ($p > 0.01$). Following is an example of a response from a control group student:

> The timely discovery of the atomic number by Moseley while studying the frequencies of x-rays of some elements. Next it was the periodic tables of Meyer and Mendeleev that were based on atomic weights and physicochemical properties of the elements (Student #4).

Responses of 9 (28 %) students from the experimental group were classified in option (a) and following is an example:

First it was the discovery of the elements themselves, next it was their atomic weights, atomic numbers and then the combination of these elements with others to form compounds. These discoveries are based on many experiments that even facilitated the prediction of new elements and compounds (Student #32).

It is interesting to observe that almost two-thirds of the control group and less than one-third of the experimental group students suggested this option. This can be attributed to the experimental treatment as these students had a more ample and comprehensive vision of the development of the periodic table. Furthermore, none of the control group students suggested options (b) (c) or (d), which is understandable, as textbooks generally ignore such factors.

Examples of Students' Responses who Suggested Option (b)

None of the control group students suggested option (b). On the other hand, 5 (16 %) students from the experimental group suggested option (b) and following is an example:

Birth of modern chemistry starting with Dalton's atomic theory, gave rise to an intense scientific activity in the 19th century. By proposing his atomic theory, Dalton helped to introduce the important concept of atomic weight, later came the contribution of Avogadro with respect to the relation between atomic weight and valence. Based on these contributions Meyer and Mendeleev elaborated their periodic tables and even predicted new elements (Student #4).

This response clearly recognizes the importance of Dalton's atomic theory which was central to the development of the atomic weights and subsequently the periodic table.

Examples of Students' Responses who Suggested Option (c)

None of the control group students suggested option (c), which dealt with the importance of the Karlsruhe Congress held in 1860, in the development of the periodic table. In contrast, 4 (13 %) students from the experimental group selected option (c) and following is an example:

One of the first factor was the recognition of the importance of atomic theory. Later the scientists started to work on atomic weights of the elements and the formation of compounds. Dalton and Gay Lussac made important contributions. However, it was Cannizaro's contribution at the Karlsruhe Congress that helped to define the concepts of atomic weight, equivalent weight, valence, etc., that eventually led the chemists to develop the different periodic laws (Student #2).

This response is a fairly good example of a historical reconstruction, in which the role of Karlsruhe Congress is presented in the context of various other developments and not just an isolated event. In this context it is interesting to analyze the concept map presented by Student #13. Besides other facets, an important aspect of this concept map (Fig. 3) is the central role that the student attributes to the Karlsruhe Congress of 1860, by establishing a direct link between

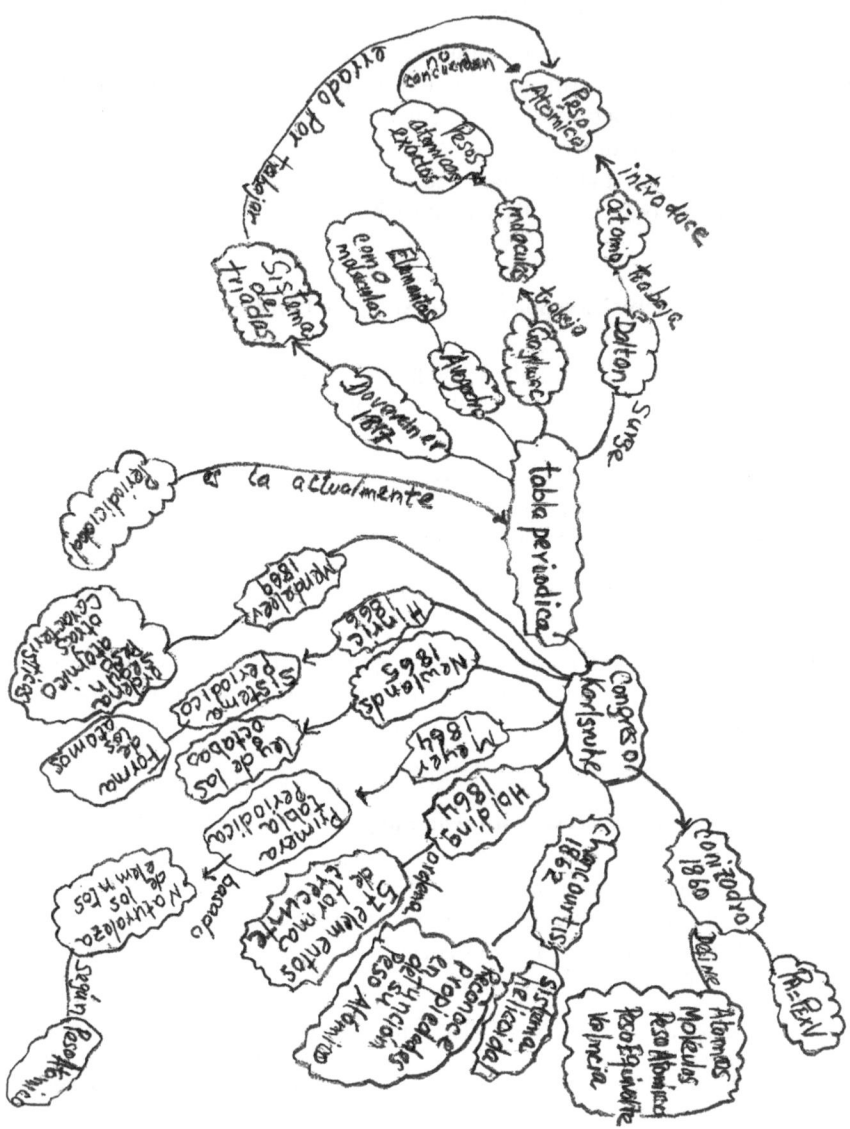

Fig. 3 Concept map drawn by Student #13. *Note* This concept map is reproduced from the student's notebook. In order to facilitate visibility all the words are retraced with pencil #2. There are some mistakes with respect to some names and properties, which were part of the original concept map

the Congress and the contributions of Mendeleev, Hinrichs, Newlands, Meyer, Odling, Chancourtois and Cannizaro. This also shows that Mendeleev formed part of a scientific community in which various participants were pursuing similar objectives. According to Gallego-Badillo et al. (2012), Karlsruhe Congress not only initiated chemistry as a scientific discipline but also provided the impetus for the beginnings of an international scientific community of chemists.

Examples of Students' Responses who Suggested Option (d)

None of the control group students suggested option (d), which dealt with the placement of the noble gases, when argon was discovered in 1895. On the other hand, 2 (6 %) students from the experimental group suggested option (d) and following is an example:

> One of the most important factors in the development of the periodic table was the prediction of new elements. Mendeleev was fully aware of this and consequently left empty spaces in the table and also corrected some of the atomic weights in order to facilitate a better placement of the elements. In this context, the discovery and placement of the inert gases (in a separate group) played a crucial role in the development and acceptance of the periodic table (Student #10).

An interesting feature of this response is that it considers the placement of the inert gases in the context of other predictions. This does make sense as the discovery and placement of argon led to the prediction of other noble gases.

Responses on this item are particularly important as it was not just selecting a particular response (multiple choice) but rather students had to suggest a factor that in their opinion was important for the development of the periodic table. All the students who suggested options (b) (c) and (d) did not simply select a particular factor but provided a background and context to their selection.

Cause of Periodicity in the Periodic Table (Students' Responses on Item 7, Posttest 2)

Responses of both groups of students on Item 7 of the Posttest 2 are presented in Table 7.

Table 7 Comparison of the performance of control and experimental group students on Item 7[a] (Posttest 2)

Response	Control ($n = 45$)	Experimental ($n = 32$)	χ^2 (Sig.)
Conceptual	–	4 (13 %)	–
Rhetorical	29 (64 %)	15 (47 %)	ns
No response	16 ((36 %)	13 (41 %)	ns

[a] Item 7: Periodicity of elements in the periodic table is: A consequence of physically observable properties (as aggregates) or chemical atoms as particles?

The idea behind this item was to evaluate students' understanding of the underlying cause of periodicity of chemical elements in the periodic table. It is plausible to suggest that both physical and chemical properties are basically a manifestation of the atomic structure and this was quite clear to Mendeleev and others, starting from the middle to the late nineteenth century. Consequently, atomic weight along with the physicochemical properties, were the two main criteria for the classification of the elements, which were in turn based on the atomic theory. Later based on the work of J. J. Thomson, G. N. Lewis, N. Bohr and H. G. J. Moseley, atomic number was found to be a better criterion that resolved many anomalies in the placement of the elements. None of the students from the control group had a conceptual response and 4 (13 %) students from the experimental group recognized the importance of the particulate nature of matter and thus responded conceptually. Following is an example of a conceptual response by a student from the experimental group:

> In the beginning the periodicity of the elements was studied by Mendeleev according to the atomic weight and physicochemical properties. Later these classifications were corrected by the valence and electron configurations [Bohr, Moseley] of the elements. At this stage it is important to clarify that the *physicochemical properties are a function of the atomic or particulate nature of the elements*, which is in turn manifested by valence and electron configuration (Student #4, italics added).

Now let us see how the conceptual response to this item differs from the rhetorical responses. Following is an example of a rhetorical response by a student from the control group:

> I consider that periodicity is a consequence of the physical properties and the chemical atoms, as the elements are ordered according to their atomic weights (Student #6).

Following is an example of a rhetorical response by a student from the experimental group:

> It is a consequence of both, as many scientists ordered the elements according to their physicochemical properties (Student #2).

It is important to note that there is a fundamental difference between the conceptual and rhetorical responses on this item. Although both types of responses emphasized the physical properties and the chemical atom, the rhetorical responses lacked the understanding that the physico–chemical properties are a consequence of the atomic nature of chemical atoms. In other words, the correct (that is conceptual) line of reasoning would emphasize the following sequence: atomic nature of the elements → physico–chemical properties → ordering of the elements. Conceptual responses explicitly followed this reasoning, which shows the importance of the atomic weight in the early periodic tables and the atomic number after the work of Moseley and others. This distinction is difficult to grasp and even many (47 %) experimental group students gave a rhetorical response. Furthermore, research in science education has shown that both high school and university freshman students have considerable difficulty in understanding the particulate nature of matter (cf. Gabel and Bunce 1994; Niaz and Montes 2012).

In order to further understand the difference between conceptual and rhetorical responses, let us consider the following responses by two experimental group students:

> The periodicity of the elements in the periodic table is a consequence of physical properties as aggregates and/or chemical properties. *Scientists ordered the elements in order to achieve periodicity* (Student #13, italics added).

> The periodicity of the elements in the periodic table is a consequence of chemical properties, such as atomic weight, valence, equivalent weight and *all these properties are linked with the atomic theory* (Student #3, italics added).

Response of Student #13 was classified as rhetorical and that of Student #3 was classified as conceptual. These two examples clearly illustrate the difference between the two types of responses. Student #13 first stated that the periodicity was a consequence of physical and chemical properties, but later asserted that periodicity was achieved because of the ordering of the elements by the scientists (see the part in italics). In contrast, Student #3 goes beyond by first invoking the physical and chemical properties and later attributing these properties to an underlying cause, namely the atomic theory (see the part in italics). Such differences between the two types of responses are important if we want our students to have a deeper understanding of the scientific enterprise.

Interviews with Experimental Group Students

Five students from the experimental group voluntarily agreed to be interviewed, which were conducted about 1 week after the students responded to Posttest 2. Each interview lasted about 45 min, which were audio-taped and later transcribed. Each interview was based on students' responses to one or more items of Posttests 1 and 2. The researcher showed the student her/his written response to a particular item and then asked relevant questions. The idea behind the interview was to explore students' thinking beyond that expressed in the written response. Following letters are used to transcribe the interviews: S = Student, R = Researcher.

Role of the Atomic Theory (First Interview, Student #3)

R: With respect to Item 1 of Posttest 1 (What was the criterion used by Mendeleev to put the elements in the established order in the periodic table?) you responded that Mendeleev placed the elements according to their atomic weight, valence, oxides, volume, and specific heat. Specifically, with respect to valence and the oxides, what are the characteristics on which these properties depend?

S: I would say it is the atomic theory.

R: This atomic theory that you are referring to is the one postulated by Dalton or the atomic theory that was postulated later.

S: I would say the one that was postulated later. When Dalton proposed his atomic theory, it had errors of measurement and he had conflicts with Gay Lussac, with respect to the volumes of the elements and the composition of the compounds. Consequently, Mendeleev's periodic table is based on the latter atomic theory.

R: Are you aware of the fact that the atomic theory used by Mendeleev emerged before the work of Thomson, Rutherford and others?

S: Yes.

R: So what was Mendeleev's basis to assert that the properties of the elements depended on the atomic theory? What helped Mendeleev to order the elements? What was there behind all this?

S: The atomic weights.

R: And the atomic weights depended on what?

S: The atomic theory.

R: Was it Dalton's or the modern atomic theory?

S: Dalton's theory.

R: Dalton had conflicts with other scientists and out of these conflicts emerged the concept of atomic weight.

S: Yes, I agree.

R: After the conflicts were resolved, Mendeleev proposed his periodic table based on his theoretical framework, and later came the discovery of the electrons, atomic nucleus, and atomic number. So what was the basis of Mendeleev's periodic table?

S: Now, I am confused.

R: Let us clarify. If Mendeleev's periodic table is based on atomic weights, and the atomic weights are based on the atomic theory, on which atomic theory did Mendeleev base his table, the modern atomic theory or the one originally proposed by Dalton?

S: For me it was the one that followed Dalton's work. As Dalton's atomic theory had many errors and he had to face conflicts, the atomic theory I am referring to is the one that emerged out of these conflicts.

R: OK. Are you suggesting that Mendeleev's periodic table was based on an atomic theory that used electrons and protons? So how is it possible that Mendeleev's periodic table emerged before the modern atomic theory? Did Mendeleev know of these particles? Did he have an idea that they existed?

S: I imagine that he did.

R: How is that possible?

S: There seems to be a contradiction and I am confused.

R: Let us discuss another point. On Item 3 of Posttest 1 (If the periodic table
     was elaborated before the modern atomic theory, how could Mendeleev and
     others do so?), you constructed a concept map in which you indicated that the
     Mendeleev's periodic table is based on the atomic theory, which is based on
     Moseley's concept of atomic number. Now, Mendeleev presented his table
     first in 1869, whereas Moseley presented his work in 1913. How do you
     explain this? What were you thinking at that moment? Did Mendeleev know
     of these concepts that were discovered by Moseley?
S: I guess that he did, as he extended the work of Dalton and others.
R: Precisely, Dalton was one of the first to propose the existence of atoms, and
     this led to the concept of atomic weight among others. Mendeleev based his
     table on these concepts, but he was not aware of the atomic number or other
     developments that led to the modern atomic theory.
S: The periodic table first emerged with ideas based on Dalton's atomic theory,
     which was not very successful. Later came Cannizaro, and both Mendeleev
     and Meyer accepted his ideas with respect to the atomic weight. This chain of
     events ended with Moseley's work with respect to atomic number based on
     the electronic structure.

Comments

Student's responses in this interview have many strands and at times he/she shows
considerable ingenuity in trying to understand a chain of events that has many
complex features. Some of the positive aspects of this student's thinking are the
following: (a) In the very first response the role of atomic theory is clearly recognized; (b) Dalton's atomic theory had to face criticisms from his contemporaries
(Gay-Lussac and others) and it was these modifications that were of help to
Mendeleev and his contemporaries; and (c) Based on the work of Cannizaro, both
Meyer and Mendeleev recognized the importance of atomic weight for the periodic table. Interestingly, both Meyer and Mendeleev attended the Karlsruhe
Congress and expressed their appreciation of Cannizaro's ideas (cf. Brito et al.
2005). Despite these positive aspects, this student had difficulties in conceptual-
izing the chain of events and she/he acknowledges the confusion. A major difficulty was the assertion (in a concept map) that the atomic theory used by
Mendeleev was based on Moseley's concept of atomic number. Interestingly,
students were not required to construct concept maps while responding to the
different exam items (Posttests # 1 and #2). This student, however, found it helpful
to understand the problem by constructing a concept map (see Fig. 4) and thus
provided interesting information with respect to the difficulties faced.

It is important to note that in the concept map, this student does seem to imply
that Mendeleev's periodic table is based on the atomic theory, which is based on
Moseley's concept of atomic number. When specifically questioned by the
researcher with respect to this relationship the student answered in the affirmative.

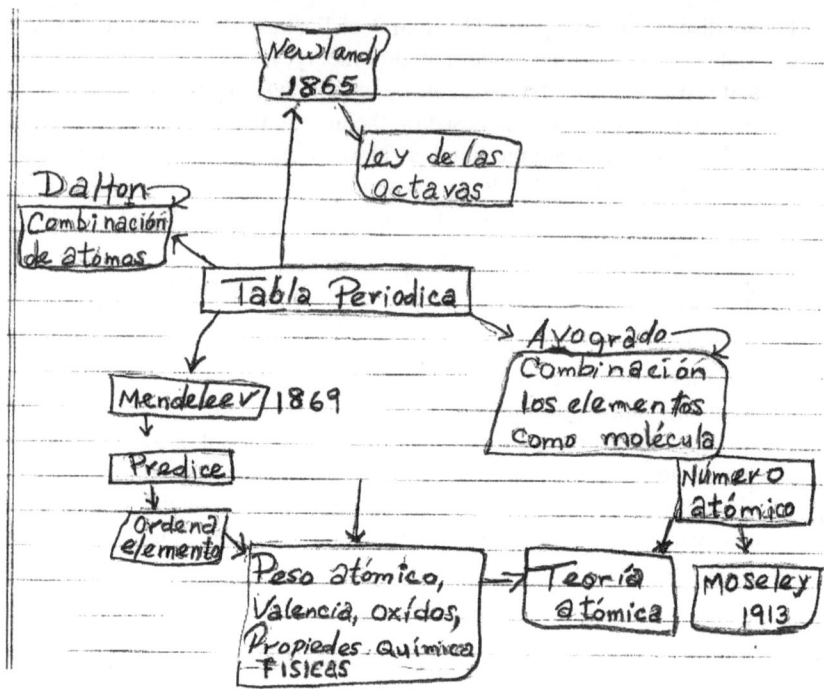

Fig. 4 Concept map drawn by Student #3. *Note* This concept map is reproduced from the student's notebook. In order to facilitate visibility all the words are retraced with pencil #2. There are some mistakes with respect to some names and properties, which were part of the original concept map

Despite the confusion, in the last response this student does seem to have captured a better understanding of the chain of events that led to changes in the periodic table based on Moseley's work related to atomic numbers. In this response, Student #3 has presented a fairly succinct historical reconstruction of the periodic table based on: Dalton's atomic theory → Cannizaro's Karlsruhe Lecture → Atomic weights → Periodic tables of Mendeleev and Meyer → Moseley → Atomic numbers → Electronic structure of the elements.

Intellectual effort involved in the work of Mendeleev and Moseley has been the subject of discussion in the chemistry education literature. Undoubtedly, the concept of atomic number developed by Moseley increased the predictive power of the periodic law. However, according to Gorin (1996), Mendeleev's contribution was more groundbreaking if we consider that: "How did Mendeleev manage to gain this insight, which today may seem commonplace but was extraordinary in its time?" (p. 490).

Progressive transitions in the history of science are frequent and hence changes in the periodic law (Mendeleev and what next) were foreseen by some scholars. Pattison Muir (1907/1975) an English chemist, writing before Moseley did his work, expressed this in the following terms: "The future will decide whether the periodic law is the long looked for goal, or only a stage in the journey, a resting-place while material is gathered for the next advances" (p. 375).

At this stage, it is important to note that experimental group students were also exposed to other sources of information, such as the textbooks and their friends in other sections of the same course that were given by different instructors. It is plausible to suggest that such interactions outside the classroom could have been the cause of confusion for some experimental group students.

Development of Better Forms of the Periodic Table (Second Interview, Student #30)

R: With respect to Item 1 of Posttest 1 (What was the criterion used by Mendeleev to put the elements in the established order in the periodic table?) you responded that Mendeleev placed the elements according to their properties. What were these properties?

S: Mostly physical properties. For example, the weight of the element changed rapidly due to oxidation.

R: This means that you consider the capacity for oxidation as a physical property. Do chemical properties have a relationship with the capacity to react of an element?

S: Yes, there is a relationship.

R: So the capacity for oxidation depends on what?

S: First, it depends on whether the element is a metal or a nonmetal. Second, it is the capacity to combine with oxygen.

R: Let us change the subject. On Item 2 of Posttest 1 (If the periodic table was elaborated before the modern atomic theory, do you think there is a relationship between the periodic table and the earlier atomic theory?), you responded yes, as this helped to explain the placement of the elements. Which atomic theory are you referring to, Dalton's atomic theory or the one that emerged after the work of Thomson, Rutherford, and others?

S: It must be the atomic theory of Dalton, as it refers to the valence electrons that helped to place the elements.

R: Let us clarify: Dalton's atomic theory postulates the existence of atoms and later the modern atomic theory is based on electrons and protons. Which atomic theory are you referring to: Dalton's atomic theory or the modern atomic theory?

S: At first, it was Dalton's atomic theory as it helped to fill the spaces in the periodic table.

R: In other words, the order of the atomic theories is justified?

S: Yes, Dalton's atomic theory helped first to order the elements in the periodic table and later the modern atomic theory confirmed it.

R: After Mendeleev's work, there were other periodic tables? Did the periodic table improve?

S: Yes the periodic table improved, as he left empty spaces for elements that were discovered later and some were not discovered. In this sense, the periodic table has improved.

Comments

First, it seems that this student had some difficulty in understanding the difference between the physical and chemical properties. Second, once again there is some ambiguity with respect to the role played by the atomic theory and the modern atomic theory. Interestingly, after the difference between the two atomic theories was clarified, the student rightly concluded that Dalton's theory helped first and later the distribution of the elements was not only confirmed by the modern atomic theory, but *facilitated better forms of the periodic table*. It is important to note that such questions are generally not only ignored by the textbooks but rather foster an inductivist vision of the elaboration of the periodic table, according to which Mendeleev had no theory or framework to guide him.

At this stage, it is important to recognize the difference between Dalton's atomic theory and its later development during the second half of the nineteenth century, before the modern atomic theory began to develop starting around 1897. Writing at beginning of the twentieth century, Merz (1904), a philosopher of science expressed this change in the following terms: "Thus the atomic theory, known to the ancients, revived by Dalton in the early years of the [nineteenth] century, and employed by chemical philosophers for half a century as a convenient symbolism, had, about the year 1860, been accepted by physicists, and used not merely as a convenient symbolism, but as a physical reality" (pp. 436–437). Despite some persistent opposition by some anti-atomists, Merz is referring to the use of the atomic theory by James Clerk Maxwell in the development of the kinetic molecular theory of gases (for details and educational implications, see Niaz 2000).

Ambiguity in Mendeleev's Writings with Respect to the Role Played by the Atomic Theory (Third Interview, Student #4)

R: With respect to Item 5 of Posttest 2 (In the acceptance of the periodic table which of the following factors was most important?) you seem to suggest that the atomic theory did not contribute toward the accommodation and prediction of the elements in the periodic table. Can you please explain?

S: Dalton's atomic theory led to the concept of atomic weight, and this is the basis of the periodic table. It contributes in the sense that in the beginning it provided the necessary order in the periodic table. However, Dalton's atomic theory does not play the same role in the actual periodic table.

R: We are referring to the acceptance of Mendeleev's periodic table.

S: Let me include Meyer here who elaborated a periodic table based on Dalton's atomic theory. Later, Mendeleev (in contrast to Meyer) based his periodic table on the chemical properties and the atomic weights. Thus, the role of the atomic theory is important but different in the periodic tables of Meyer and Mendeleev.

R: In that period of time, Dalton's atomic theory was the subject of controversy in the scientific community.

S: Among those involved in this controversy was Meyer, when he proposed his periodic table.

R: Let us recall that Dalton's atomic theory proposed the existence of atoms, which combine to form compounds. What happened to Mendeleev's periodic table? Why did Meyer's periodic table not have the same acceptance as that of Mendeleev's. Although at times Mendeleev was ambiguous with respect to the role played by the atomic theory, he was generally categorical with respect to its importance and following is an example: "The periodic law has clearly shown that the masses of the atoms increase abruptly by steps, which are clearly connected in some way with Dalton's law of multiple proportions … the theory of the chemical elements with Dalton's theory of multiple proportions, or atomic structure of bodies, the periodic law opened for natural philosophy a new and wide field for speculation" (Mendeleev 1889, p. 642).

S: This is what has confused me as on reading some statements by Mendeleev I came to the conclusion that he was not in agreement with Dalton's atomic theory.

R: With respect to Item 6 of Posttest 2 (Which factors were important for the development of the periodic table?) you mentioned the following factors: Contributions of Dalton and Avogadro, Karlsruhe Congress, contributions of Meyer and Mendeleev. Please remember that as compared to the previous item, in this case you could generate your own factors. At this stage, I invite you to reconsider your response and if necessary add any other factor.

S: Yes, I would like to refer to two contributions before Mendeleev. One was by Odling who organized 57 elements according to their atomic weights, and the other was Newland's Law of Octaves, according to which the properties of the elements repeated after every seven elements. An important feature of all these contributions was that when the elements were placed in groups, these had similar properties. Another factor was the controversy related to the placement of gas Argon. It was found that its properties and atomic weight did not permit its placement in either the group of halogens or alkali metals. The problem was solved by placing these inert gases in a new group between the halogens and the alkali metals. After this came, the discovery of the electron and subsequently the atomic number.

R: Now, on Item 7 of Posttest 2 (Periodicity of elements in the periodic table is: a consequence of physically observable properties or chemical atoms as particles?) you responded: "In the beginning the elements in the periodic table were accommodated by Mendeleev according to their density, atomic weight, and oxidation group, that is both chemical and physical properties. These were later reorganized based on corrections of their chemical properties especially with respect to the concept of valence and electron configuration …". Can you explain your ideas, as in Item 6 (Posttest 2) you mentioned that the concept of valence was defined at the Karlsruhe Congress, while in this response and also on Item 5 (Posttest 2) you mentioned that the concept of valence was defined after the discovery of the electrons and the electron configurations?

S: Yes, that is correct. Let me explain: what I wanted to say is that the role played by valence after the Karlsruhe Congress is the same as that of electron configurations in actual practice.

R: However, we are referring to the period in which Mendeleev's periodic table emerged. When Mendeleev presented his table he was not aware of electrons and much less of electron configurations. It is for this reason that I think that you do not clearly differentiate between Dalton's atomic theory and the modern atomic theory. In your response to Item 7 (Posttest 2), you further stated that: "It is recommended that this point be clarified. For example, when we study the density of an element (physical property) we take it as an aggregate property. However, it is the chemical properties that are most relevant, as these are determined by the atomic nature of particles". Which particles are you referring to?

S: I am referring to the atoms in the chemical atoms and not their mass as part of a compound.

R: How did Mendeleev know of the existence of these particles?

S: Through the atomic theory of Dalton, as this led him to determine the atomic weights and other properties.

Comments

There are various interesting features (at times quite complex) in the responses of this student and following are some relevant examples:

(a) Role played by the atomic theory in the periodic tables of Mendeleev and Meyer.

(b) Ambiguity in Mendeleev's writings with respect to the role played by the atomic theory. This is discussed in the rationale section and has been the subject of discussion in the history and philosophy of science literature (Christie and Christie 2003). Due to the positivist milieu of his time, Mendeleev also emphasized the importance of inductive generalizations, and hence the ambiguity. It is a positive aspect of this study that this student specifically referred to this ambiguity, which was discussed in class.

(c) It is interesting to note that this student understood that the concept of valence as used by Mendeleev was clarified by Cannizaro at the Karlsruhe Congress. Furthermore, the student establishes a 'historical analogy' by suggesting that the role played by valence after the Karlsruhe Congress is similar to that of electron configurations in actual practice.

(d) The interview provided the researcher (with the active participation of the student) to collate a wide range issues based on student's responses on different items, such as Items 5, 6, and 7 of Posttest 2. These items in turn were elaborated with the objective of constituting a sequence of events and the ensuing concepts.

Role of Accommodations and Predictions (Fourth Interview, Student #8)

R: On Item 5 of Posttest 2 (In the acceptance of the periodic table, which of the following factors was most important: Accommodation, Prediction, and Correction of atomic weights?), you responded that all three factors were important. However, you also stated that correction and accommodation were more important. In what did you base your response?

S: Well, accommodation of the elements is important as it helps to order the elements based on the properties of the individual atom and the atomic weights. In the case of predictions, Mendeleev used to leave empty spaces and predicted the properties of the elements that were discovered later. When the new elements were discovered and their properties coincided with those predicted by Mendeleev, it provided greater validity to the periodic table.

R: On Item 6 of Posttest 2 (Which factors were important for the development of the periodic table?), you seem to suggest the same factors as provided in Item 5. Do you realize that on Item 6 you could suggest factors of your own preference. At this stage would you like to suggest some other factor? For example, just consider: What was the fundamental fact that facilitated the appearance of the periodic table?

S: Let us see: For the appearance of the periodic table, the discovery of the elements was perhaps very important.

R: On Item 7 of Posttest 2 (Periodicity of the elements in the periodic table is a consequence of: physically observable properties or chemical atoms as particles?), you responded that it depends on atomic weights, atomic numbers, electronegativity, and among other properties. Of course, you are aware that atomic numbers and electronegativity were postulated after Mendeleev, by the modern atomic theory.

S: Yes, Mendeleev based his table primarily on the atomic weights.

Comments

An important facet of this interview is that this student related the prediction of new elements by Mendeleev and the coincidence in the properties of the two (predicted and the actual elements after discovery) as a source of greater validity. According to a major historian of the periodic table, after the discovery of gallium in 1875, "…Mendeleev rightly concluded that the validity of the periodic system of elements could no longer be questioned. The confirmation of this prediction may certainly be called the culminating point in the history of the periodic system" (Van Spronsen 1969, p. 221). This continues to be a point of contention among philosophers of science, namely what made Mendeleev's law valid—accommodations dating from 1869 or the predictions from 1875 onwards. For a recent debate on this issue see Lipton (2005a, b).

Relationship Between Dalton's Atomic Theory and the Modern Atomic Theory (Fifth Interview, Student #9)

R: Item 5 of Posttest 2 refers to three fundamental factors responsible for the acceptance of the periodic table, namely accommodation of elements, prediction of new elements, and correction of atomic weights. In your response you indicated that all three factors were equally important. Would you like to add anything to your response?

S: In my opinion, all three factors are not only important but also interrelated.

R: Item 6 of Posttest 2 asked: Which factors were important for the development of the periodic table and you could generate factors of your own choice. Would you like to add anything to your response?

S: I would like to emphasize the discovery of the noble gases starting with argon that required the inclusion of a new group in the periodic table. Thomson's postulation of the electron was also very important as it eventually led to the formulation of electron configurations that modified the placement of various elements in Mendeleev's periodic table.

R: Item 7 of Posttest 2 asked: Periodicity of elements in the periodic table is: a consequence of physically observable properties or chemical atoms as particles? You responded: "Periodicity of elements in the periodic table is a consequence of chemical properties as the elements were placed according to their electronegativity, oxidation state …"

S: Actually this took place later. In the beginning elements were accommodated according to their density, atomic weights, etc.

R: In the beginning when the elements were accommodated based on these properties, was there a relationship between these properties and the atomic theory?

S: Yes, there was a relationship because the atomic theory led to the postulation of atomic weights. In the early periodic tables elements were placed according to their atomic weights and the capacity to combine with oxygen [cf. examples provided by Mendeleev 1889, pp. 640–641].

R: Is there a relationship between atomic weights and Dalton's atomic theory?

S: Yes, and this was the main reason why many scientists started to elaborate the periodic table based on atomic weights.

R: And these atomic weights depend on what?

S: Atomic weights: Do they have to depend on something?

R: In other words, when Dalton proposed the atomic weights he must have sustained and looked for support in some basic concepts.

S: Oh yes, first in the study of the atoms which led to the idea of atomic weights.

R: Is there a relationship between Dalton's atomic theory and some of the factors that you mentioned, such as electronegativity and the state of oxidation.

S: No, as Dalton was not aware of the existence of electrons, which were postulated later by Thomson.

R: Does this refer only to electronegativity or it also refers to the state of oxidation?

S: It was difficult for Dalton to understand the states of oxidation as he considered the atoms to be indivisible.

R: So you are referring to Dalton's atomic theory and not the modern atomic theory?

S: Yes. However, there is a relationship between the two, as new aspects of atomic structure were discovered, such as electrons.

Comments

It is important to note that this student (#9) had already provided conceptual responses on Items 2 and 3 of Posttest 1. As the Posttest 2 was administered about a month later, it was a good opportunity to follow the retention of some of the concepts already referred to and used by this student in Posttest 1. Following are some of the salient features of this interview:

(a) On Items 5 and 6 of the Posttest 2, this student provided fairly good responses that were classified as conceptual. On Item 6 of Posttest 2, when asked: Would you like to add anything to your response? This student referred to the inclusion of argon in the periodic table, which caused problems for the validity of the periodic table. Interestingly, this was also suggested in her/his response to Item 3 of Posttest 1. It is plausible to suggest that the student recalled this information in a new context.

(b) On Item 7 of Posttest 2, this student to begin with did not refer to Dalton's atomic theory as a cause of periodicity. However, as the interview continued and under the guidance of the researcher, she/he did recognize the importance of *Dalton's atomic theory and its relationship with the modern atomic theory*. Apparently, the student did not recall the relevant information, as on Item 2 of Posttest 1, she/he had recognized the importance of Dalton's atomic theory for the periodic table.

(c) On Item 7 of Posttest 2, this student referred to the capacity of oxygen to combine with the elements as a criterion for placing the elements in the periodic table. Interestingly, Mendeleev (1889, pp. 640–641) himself referred to the increasing quantity of oxygen in the following oxides as evidence for the role played by the atomic theory in the elaboration of the periodic table: Ag_2O, Cd_2O_2, In_2O_3, Sn_2O_4, Sb_2O_5, Te_2O_6, and I_2O_7.

Conclusions and Educational Implications

Based on the theoretical rationale and the four research questions, a major objective of this study is to facilitate students' conceptual understanding of the development of the periodic table. The seven evaluation items included in Posttests 1 and 2 were formulated in order to constitute a sequence of progressive transitions in understanding the periodic table. The idea of progressive transitions is taken from the history of science (Lakatos 1970) and its application in science education (Niaz 2001, 2009a, b). The fundamental assumption behind this sequence was that the interaction with the items itself could be thought provoking, and thus provide the students an opportunity to reason and reconsider their understanding. Item 1 was fairly general and referred to the criterion used by Mendeleev to classify the elements. Item 2 was more specific and referred to a possible relationship between the periodic table and the earlier atomic theory (Dalton and others). Item 3 was much more specific and referred to how could Mendeleev and others construct the periodic table before the modern atomic theory was formulated. Item 4 was even more specific and referred to the origin of the idea of ordering the elements, namely there was a scientific community interested in the subject. Starting with Posttest 2 (Items 5, 6 and 7), students were required to select factors that played an important role in the acceptance of the periodic table. Item 5 asked students to select one of the following factors: Accommodation of the elements, prediction of elements that were discovered later and corrections of atomic weights. Item 6, again referred to the factors that were important for the development of the periodic table, with the difference that the students could provide their own list of factors, and thus interact with science content. Finally, Item 7 asked the crucial question, what was the cause of periodicity? It is plausible to suggest that on Items 1, 2, 3, and 4, which formed part of

Posttest 1, students could go back and forth and thus revise their response. The same could have occurred on Items 5, 6, and 7 which formed part of Posttest 2. The closely structured sequence of items and the opportunity to revise the responses provided important feedback to the students.

It is plausible to suggest that this study helped to facilitate a greater understanding of the four research questions that helped to guide the study. Students responses on all items (especially Items 1, 2, 3, 4, and 7) and the interviews provided insight with respect to Research question #1 (Role played by Dalton's atomic theory in the origin and development of the periodic table) and Research question #2 (Role of periodicity as a function of atomic theory). Items 5 and 6 provided evidence for Research question #3 (Prediction of new chemical elements). On Item 5, where students were provided with a list of factors that were important for the acceptance of the periodic table, approximately 14 % of the experimental group students selected factors that provided some insight with respect to the research question. Similarly, on Item 6, where the students generated their own factors approximately 16 % of the experimental group students provided some insight with respect to the research question. It seems that the evidence for this research question (#3) was not conclusive as the students could combine factors, and hence it was difficult to isolate the importance of each factor. Evidence for research question #4 was provided by all the items and the interviews and a detailed eight point design of a teaching strategy is included in this section.

Conceptual Responses by Experimental Group Students

Experimental group students provided conceptual responses on all items. Item 1 dealt with atomic theory as the criterion used by Mendeleev to order the elements, and 19 % of the students responded conceptually. Item 2 dealt with the relationship between the periodic table and the early atomic theory, and 47 % of the students responded conceptually. Item 3 dealt with the question as to how Mendeleev could elaborate the periodic table before the modern atomic theory, and 28 % of the students responded conceptually. Item 4 asked if the idea of ordering the elements originate with Mendeleev, and 13 % of the students responded conceptually. Item 7 referred to periodicity as a function of the chemical atoms (atomic theory) and 13 % of the students responded conceptually. Apparently, Items 1 and 7 refer to the same conceptual aspect and still percentage of students responding conceptually decreased from 19 to 13 %. It seems that "periodicity as a function of chemical atoms" in Item 7 was more difficult to understand than "criterion used by Mendeleev to order the elements" in Item 1.

Items 5 and 6 were slightly different in which responses were not classified as conceptual or rhetorical. Nevertheless, a comparison of performance on Items 5 and 6 provides interesting insight into students' thinking and understanding. On Item 5, 34 % of experimental group students selected option (a), that is accommodation of the elements, and 28 % selected option (c), that is corrections of atomic weights. In contrast, on Item 6 (see Table 6), where the students could provide their own factors, entirely new options appeared, such as: (b) Dalton's atomic theory (16 %); (c) Karlsruhe Congress (13 %); and (d) Placement of the noble gases (6 %), which gives a total of 35 %. This clearly shows that given the opportunity, experimental group students can go beyond the factors discussed in the traditional classrooms and textbooks. It is important to note that the experimental treatment in this study (see Rationale section) explicitly dealt with the role played by the following in the development of the periodic table: Dalton's atomic theory (b), Karlsruhe Congress (c), and Placement of the noble gases (d). Interestingly, none of the control group students provided responses that could be classified as options (b), (c), and (d) of Item 6, which clearly shows the difference in the understanding of the two groups of students in this study.

Conceptual Responses by Control Group Students

As control group students were not exposed to the experimental treatment, it was not expected that they would respond conceptually. Nevertheless, one student on Item 1, two students on Item 2 and one student on Item 3 responded conceptually. How can we explain conceptual responses by control group students who received instruction in a previous semester, and hence could not have interacted with experimental group students? In order to respond to this question let us analyze the response provided by the control group student (#33) on Item 2, which has the following critical aspects: (a) A clear distinction between the early atomic theory and the modern atomic theory; (b) Some properties of the atoms were known quite early; (c) Early chemists must have had some notions of the atomic theory based on properties of the atoms; (d) Study of the atoms led to the study of the physical and chemical properties of the elements and subsequently their ordering in the periodic table. All these four critical aspects are in general discussed in almost all general chemistry textbooks, namely Dalton's atomic theory, physical and chemical properties of the elements and their compounds, contributions of Gay-Lussac, Avogadro, and others, and of course the early attempts to order the elements as early as 1817 by Döbereiner. It is plausible to suggest that given the opportunity to reflect and appropriate test format (as provided by this study, through Items 1–7), at least some students can establish a relationship between the early atomic theory, properties of the elements and their ordering in a periodic table. Conceptual responses by control group students provide a good argument for including such material in classroom discussions, especially in the context of the periodic table.

Multiple Data Sources

Findings in this study are supported by the following multiple data sources: (a) Responses from control and experimental group students on seven items that formed part of Posttests 1 and 2. Items 1–4, 6, and 7 were open-ended and the students were not constrained by the test format. Only on Item 5, students had to select an option and even in this item they could combine different options according to their convenience. Format for Item 5 was so selected in order to compare students' responses on Item 6, in which they suggested their own factors that they considered important for the development of the periodic table. (b) Concept maps constructed by the students before and after Posttest 1 were particularly helpful in facilitating understanding. Furthermore, before constructing concept maps for the second time (after Posttest 1), the previous concept maps were discussed in small groups among the students and the instructor. (c) Semi-structured interviews with five voluntary students provided greater insight into students' written responses on Posttests 1 and 2 and also gave the students the possibility to add new information. Comparison of the responses on test items, concept maps and the interviews provided considerable depth to the findings of this study. Working with multiple data sources approximates to *triangulation of data sources* and has been endorsed by Johnson and Onwuegbuzie (2004): "Researchers should collect multiple data using different strategies, approaches and methods in such a way that the resulting mixture or combination is likely to result in complementary strengths" (p. 18).

Design of Teaching Strategies

As an educational implication of this study (Research question #4), we suggest that while teaching the periodic table both at the high school and introductory university courses the following can constitute guiding principles:

1. How could a simple arrangement of the elements based on atomic mass (atomic weight for Mendeleev) provided such regularities as observed in the periodic table?
2. Many scientists including Mendeleev were continually trying to understand the underlying reason for periodicity. These efforts went through various tentative attempts to understand and classify the elements. On the contrary, most text-books give the impression that for almost 100 years (1820–1920), scientists had no idea or never asked the question as to whether there could be an underlying rationale for explaining periodicity. Furthermore, textbooks in general ignore the tentative nature of scientific knowledge (for details, see Niaz and Maza 2011).

3. Besides Mendeleev in 1869, following co-discoverers of the periodic table also made important contributions: De Chancourtois in 1862, Odling in 1864, Meyer in 1864, Newlands in 1865, and Hinrichs in 1866.

4. Even before the modern atomic theory (starting 1897) scientists were well aware that periodicity in the periodic table is a function of the atomic theory.

5. Accommodations and predictions of elements provided important evidence for the acceptance of the periodic law and it would be helpful to emphasize both in the classroom.

6. Based on a historical reconstruction, the following sequence of heuristic principles can help to facilitate understanding: Accumulation of atomic weights of the elements in the early nineteenth century, Attempts to classify elements starting in 1817, Karlsruhe congress in 1860, Cannizaro's contributions, Mendeleev's first periodic table in 1869, Corrections of known atomic weights, Discovery of argon in 1895, and Contribution of Moseley in 1913 that led to the periodic table being based on atomic numbers.

7. Implementation of these guiding principles constitutes what in the history of science and science education literature has been referred to as "science in the making" (for details, see Niaz 2012).

8. An effective way in which to bridge the gap between how we teach science (periodic table in this case) and what scientists actually do, that is "science in the making" is through the inclusion of humanizing aspects of the history of science in the form of a story (contextual teaching). Klassen (2006) has referred to this contextual approach in cogent terms: "School science lacks the vitality of investigation, discovery, and creative invention that often accompanies *science-in-the-making* … The humanizing and clarifying influence of history of science brings the science to life and enables the student to construct relationship that would have been impossible in the traditional decontextualized manner in which science has been taught" (p. 48, emphasis added).

Comparing the Research Methodologies of Newton and Mendeleev

In the rationale section, we referred to some ambivalence in Mendeleev's thinking with respect to the periodic table being an induction from verified facts or based on a hypothesis according to which periodicity was a function of the atomic theory. In order to facilitate a better understanding of Mendeleev's research methodology, it would be helpful to compare it with that used by Newton (cf. Erduran 2007). In his public statements, Newton stated categorically "hypotheses non fingo" (I do not feign hypotheses). This leads to a dilemma: Did Newton formulate his law of gravitation based entirely on experimental observations, as he himself claimed. If the answer is in the affirmative, then he should have been aware that charged bodies would not follow the law of gravitation. According to Giere (1999), the idea of charged bodies was included in the physicists' agenda almost a century after Newton. Similarly, Kuhn (1977) considered the "Newtonian method" to be more of an idealization and that in order to formulate his law of gravitation, Newton inevitably resorted to the elaboration of a hypothesis. Duhem (1914), a philosopher of science and chemist was particularly critical of Newton and considered that the "Newtonian method" attractive as it may appear, was a dream. This clearly shows that when faced with difficulties scientists often resort to 'speculation' and formulate hypotheses for which they may not have convincing experimental evidence. Indeed, it would be much more motivating to students if we teach science as practiced by scientists (Niaz 2010). Inclusion of such methodological aspects in the classroom can help students to understand not only the periodic table but also scientific research methodology in general.

References

Atkins, P., & Jones, L. (2002). *Chemical principles: The quest for insight* (2nd ed.). New York: Freenman.

Atkins, P., & Jones, L. (2008). *Chemical principles: The quest for insight* (4th ed.). New York: Freeman.

Bensaude-Vincent, B. (1986). Mendeleev's periodic system of chemical elements. *British Journal for the History of Science, 19*, 3–17.

Ben-Zvi, N., & Genut, S. (1998). Uses and limitations of scientific models: The periodic table as an inductive tool. *International Journal of Science Education, 20*(3), 351–360.

Bohr, N. (1913). On the constitution of atoms and molecules, Part I. *Philosophical Magazine, 26* (Series 6), 1–25.

Brady, J. E., Russell, J. W., & Holum, J. R. (2000). *Chemistry: Matter and its changes.* New York: Wiley.

Brito, A., Rodríguez, M. A., & Niaz, M. (2005). A reconstruction of development of the periodic table based on history and philosophy of science and its implications for general chemistry textbooks. *Journal of Research in Science Teaching, 42*, 84–111.

Brush, S. G. (1996). The reception of Mendeleev's periodic law in America and Britain. *Isis, 87*, 595–628.

Brush, S. G. (2007). Predictivism and the periodic table. *Studies in History and Philosophy of Science, 38*, 256–259.

Chang, R. (2007). *Chemistry* (9th ed.). New York: McGraw-Hill.

Christie, J. R., & Christie, M. (2003). Chemical laws and theories: A response to Vihalemm. *Foundations of Chemistry, 5*, 165–174.

De Milt, C. (1951). The congress at Karlsruhe. *Journal of Chemical Education, 28*, 421–425.

Duhem, P. (1914). *The aim and structure of physical theory* (second edition, trans. Philip P. Wiener, first published 1906 as *La théorie physique: Son objet, sa structure* by Marcel Rivière Cie, Paris. First English translation by Princeton University Press, 1954 and later by Atheneum in 1962. The book was originally published as a series of articles in French in the years 1904–1905). New York: Atheneum.

Erduran, S. (2007). Breaking the law: Promoting domain-specificity in chemical education in the context of arguing about the periodic law. *Foundations of Chemistry, 9*, 247–263.

Gabel, D. L., & Bunce, D. M. (1994). Research on problem solving: Chemistry. In D. L. Gabel (Ed.), *Handbook of research on science teaching and learning.* New York: Macmillan.

Gallego-Badillo, R., Gallego-Torres, A. P., & Pérez-Miranda, R. (2012). El Congreso de Karlsruhe: Los inicios de una comunidad científica. *Educación Química, 23*, 280–289.

Giere, R. N. (1999). *Science without laws.* Chicago: University of Chicago Press.

Gordin, M. D. (2004). *A well-ordered thing: Dmitrii Mendeleev and the shadow of the periodic table*. New York: Basic Books.

Gorin, G. (1996). Mendeleev and Moseley: The principal discoverers of the periodic law. *Journal of Chemical Education, 73*(6), 490–493.

Holtzclaw, H. F., & Robinson, W. R. (1988). *General chemistry* (8th ed.). Lexington, MA: Heath.

Johnson, R. B., & Onwuegbuzie, A. J. (2004). Mixed methods research: A research paradigm whose time has come. *Educational Researcher, 33*, 14–26.

Klassen, S. (2006). A theoretical framework for contextual science teaching. *Interchange, 37*, 31–62.

Kuhn, T. S. (1977). The function of measurement in modern physical science. In *Essential tension* (pp. 178-224). Chicago: University of Chicago Press. (originally published in *Isis, 52*, 161–190, 1961).

Lakatos, I. (1970). Falsification and the methodology of scientific research programmes. In I. Lakatos & A. Musgrave (Eds.), *Criticism and the growth of knowledge* (pp. 91–195). Cambridge: Cambridge University Press.

Lipton, P. (2005a). Testing hypotheses: Prediction and prejudice. *Science, 307*(14 January), 219–221.

Lipton, P. (2005b). Response. *Science, 308* (3 June), 1411–1412.

Mahan, B., & Myers, R. J. (1990). *University chemistry* (4th ed., Spanish). Menlo Park, CA: Benjamin Cummings.

McMurry, J., & Fay, R. C. (2001). *Chemistry* (3rd ed.). Upper Saddle River, NJ: Prentice Hall.

Mendeleev, D. (1879). The periodic law of the chemical elements. *The Chemical News, 40*, No. 1042.

Mendeleev, D. (1889). The periodic law of the chemical elements. *Journal of the Chemical Society, 55*, 634–656 (Faraday lecture, delivered on 4 June 1889).

Mendeleev, D. (1897). *The principles of chemistry (2nd English ed., trans of 6th Russian ed.)*. New York: American Home Library Company.

Merz, J. T. (1904). *A history of European thought in the nineteenth century* (Vol. 1). London: William Blackwood & Sons.

Moore, J. W. (2003). Editorial: Turning the (Periodic) tables. *Journal of Chemical Education, 80*(8), 847.

Moore, J. W., Stanitski, C. L., & Jurs, P. C. (2002). *Chemistry: The molecular science*. Orlando, FL: Harcourt College.

Moseley, H. G. J. (1913a). High frequency spectra of the elements. *Philosophical Magazine, 26*, 1025–1034.

Moseley, H. G. J. (1913–1914). Atomic models and X-ray spectra. *Nature, 92*, 554.

Niaz, M. (2000). A rational reconstruction of the kinetic molecular theory of gases based on history and philosophy of science and its implications for chemistry textbooks. *Instructional Science, 28*(1), 23–50.

Niaz, M. (2001). Understanding nature of science as progressive transitions in heuristic principles. *Science Education, 85*, 684–690.

Niaz, M. (2009a). *Critical appraisal of physical science as a human enterprise: Dynamics of scientific progress*. Dordrecht, The Netherlands: Springer.

Niaz, M. (2009b). Progressive transitions in chemistry teachers' understanding of nature of science based on historical controversies. *Science & Education, 18*, 43–65.

Niaz, M. (2010). Are we teaching science as practiced by scientists? *American Journal of Physics, 78*(1), 5–6.

Niaz, M. (2012). *From 'science in the making' to understanding the nature of science: An overview for science educators*. New York: Routledge.

Niaz, M., Aguilera, D., Maza, A., & Liendo, G. (2002). Arguments, contradictions, resistances, and conceptual change in students' understanding of atomic structure. *Science Education, 86*, 505–525.

Niaz, M., & Maza, A. (2011). *Nature of science in general chemistry textbooks*. Dordrecht, The Netherlands: SpringerBriefs in Education.

Niaz, M., & Montes, L. A. (2012). Understanding stoichiometry: Towards a history and philosophy of chemistry. *Educación Química, 23*, 290–297.

Niaz, M., Rodríguez, M. A., & Brito, A. (2004). An appraisal of Mendeleev's contribution to the development of the periodic table. *Studies in History and Philosophy of Science, 35*, 271–282.

Novak, J. D. (1990). Concept mapping: A useful tool for science education. *Journal of Research in Science Teaching, 27*(10), 937–949.

Pattison Muir, M. M. (1887). On the teaching of chemistry. *Nature, 36*, 536–538.

Pattison Muir, M.M. (1907/1975). *A history of chemical theories and laws.* New York: Arno Press (First published by Wiley in 1907).

Rammelsberg, C. F. (1874). *Grundriss de chemie gemäss den neueren ansichten* (4th ed.). Berlin: Habel.

Ramsay, W. (1897). An undiscovered gas (address to the Section of Chemical Sciences of the British Association). *Nature, 56*, 378–382.

Reger, D. L., Goode, S. R., & Mercer, E. (1997). *Chemistry: Principles and practice.* Philadelphia: Saunders.

Robinson, J. (2000). The paradigm changes—but do our students know that? *Journal of College Science Teaching, 29*, 177–182.

Russo, S., & Silver, M. (2002). *Introductory chemistry* (2nd ed.). San Francisco: Benjamin Cummings.

Rutherford, E. (1911). The scattering of alpha and beta particles by matter and the structure of the atom. *Philosophical Magazine, 21*, 669–688.

Shapere, D. (1977). Scientific theories and their domains. In F. Suppe (Ed.), *The structure of scientific theories* (2nd ed., pp. 518–565). Chicago: University of Illinois Press.

Sisler, H. H., Dresdner, R. D., & Mooney, W. T. (1980). *Chemistry: A systematic approach.* New York: Oxford University Press.

Thomson, J. J. (1897). Cathode rays. *Philosophical Magazine, 44*, 293–316.

Van Spronsen, J. (1969). *The periodic system of chemical elements. A history of the first hundred years.* Amsterdam: Elsevier.

Wartofsky, M. W. (1968). *Conceptual foundations of scientific thought: An introduction to the philosophy of science.* New York: Macmillan.

Weisberg, M. (2007). Who is a modeler? *British Journal for the Philosophy of Science, 58*, 207–233.